The Plumber's Toolbox Manual

The Plumber's Toolbox Manual

Louis J. Mahieu

ARCO

New York

TAB BOOKS Inc. offers software for sale. For information and a catalog, please contact TAB Software Department, Blue Ridge Summit, PA 17294-0850.

ARCO

Simon & Schuster, Inc.
Gulf+Western Building
One Gulf+Western Plaza
New York, NY 10023

DISTRIBUTED BY PRENTICE HALL TRADE

Manufactured in the United States of America

1 2 3 4 5 6 7 8 9 10

Library of Congress Cataloging-in-Publication Data

Mahieu, Louis J.
The plumber's toolbox manual / Louis J. Mahieu : In Plain English, Inc.
p. cm.
Includes index.
ISBN 0-13-683806-5
1. Plumbing—Equipment and supplies—Handbooks, manuals, etc. 2. Plumbing—Handbooks, manuals, etc. I. In Plain English Inc. II. Title.
TH6299.M34 1989
696'.1—dc19 89-35
CIP

FIGURE CREDITS

Figures 1-2, 1-4, 1-5, 1-8, 1-13, 1-14, 1-15, 1-19, 1-22, 1-29, 1-32, 1-34, 1-35, 1-36 courtesy of Stanley Tools, a division of The Stanley Works.

Figures 1-3, 1-6, 1-9, 1-10, 1-17, 1-23, 1-24, 1-25, 1-26, 1-27, 1-28, 1-30, 1-37, 1-38, 4-4, 4-10, 4-11, courtesy of The Ridge Tool Company.

Figure 1-33 courtesy of The Milwaukee Electric Tool Company.

Figures 4-14, 4-16 courtesy of NIBCO, Inc.

Figures 4-20, 4-21, 4-23, 4-24, courtesy of Price Pfister.

Figures 4-22, 4-25, 4-28, 4-29 courtesy of American Standard.

This book is dedicated to the memory of
my father, Charles J. Mahieu, Sr.

Part One—Tools and Equipment of the Trade

1 Tools and Equipment / 1

Air-Acetylene Equipment / 2
Bars, Steel / 5
Benders, Copper Tubing / 6
Brace, Standard / 7
Bits for Brace / 8
Cast Iron Assembly Tool / 9
Caulking Irons / 10
Chisels / 11
Copper Cleaning Tools / 12
Drain Cleaning Equipment / 14
Drills, Electric / 16
Drills, Hand / 19
Drive Sockets / 20
Files / 23
Flaring Tools / 24
Hammers / 25
Ladders / 26
Lead Joint Equipment / 27
Levels / 29
Nut Drivers / 30
Oilers / 30
Pipe Cutters, Steel / 31
Pipe Cutters, Cast-Iron / 32
Pipe Extractor / 33
Pipe Tap / 34
Pipe Threaders, Manual / 34
Pipe Threaders, Electric / 36
Pipe Vises / 37
Plastic-Pipe Tools / 37
Pliers / 38
Plumb Bob / 39
Reaming Tools / 39
Rules and Tapes / 40
Safety Equipment / 41
Saws, Hand / 41
Saw, Power / 44
Screwdrivers / 45
Snips / 46
Squares / 47
Swaging Tool / 47
Testing Equipment / 48
Tubing Cutters / 49
Wrenches, General Purpose / 51
Wrenches, Pipe / 55
Wrenches, Repair / 57

2 Pipes and Fittings / 61

Pipes / 61
Pipe Fittings / 77
Faucets and Valves / 111

Part Two—Standard Plumbing Procedures

3 Preventive Maintenance and Troubleshooting / 115

Appliances / 116
Fixtures / 127
Leaks / 131
Noises in the Piping System / 135
Pipe Stoppages / 136
Potable Water Systems / 137
Sewage Handling Systems / 138
Weather-Caused Problems / 141

4 The ABCS of the Trade / 143

Basic Plumbing Systems / 143
Cutting and Joining Pipe / 149
Installation Practices / 163
Potable Water Service / 176
Repair Techniques / 180
Roughing In / 183

Part Three—Plumbing Fundamentals

5 Basic Mathematics and Formulas / 187

Definitions / 187
Basic Arithmetic / 190

Geometry / 196
Trigonometry / 207
Formulas Often Used by Plumbers / 207

6 Measurement and Conversion / 216

Using a Ruler / 216
Metric System / 220

7 How to Read Blueprints / 225

GLOSSARY / 237

Part Four—Appendices

APPENDIX 1: Rules of Safety / 267
APPENDIX 2: Codes and Regulations / 272
APPENDIX 3: Useful Tables / 274
APPENDIX 4: Commonly Used Abbreviations / 301
INDEX / 303

PART ONE

TOOLS AND EQUIPMENT OF THE TRADE

1

Tools and Equipment

Today's plumber needs versatile, durable, and modern hand tools. Not long ago hand tools were generally standardized; they were used, with minor changes, year in and year out. They were used to join cast-iron soil pipe; to cut, ream, and thread brass and steel pipes; and to join (wipe) lead pipes.

Times change, and so does technology. A few years ago wiping lead joints was a necessary and demanding skill of the trade. Today, it is fast becoming an obsolete art form, used only in maintenance and repair work. Another example of change in the industry is the joining of cast-iron soil pipe and fittings. For many years plumbers were allowed to use only lead and oakum to join cast-iron soil pipe. Now they can use (along with lead and oakum) neoprene gaskets and, lately, a neoprene sleeve with a stainless-steel jacket and clamp.

To make lead and oakum joints, you needed equipment to melt and pour lead. You needed ball peen hammers, packing irons, steel chisels, and other tools. Now, to joint cast-iron soil pipe using neoprene gaskets, a joining tool and lubricant are used. To join cast-iron soil pipe and fittings with a neoprene and stainless-steel jacket, one tool—a torque wrench—is used. Today's professionals must keep up with such changes in technology and with the development of new hand tools.

Ways to keep up with these changes include reading trade journals, joining professional trade organizations, contacting manufacturers, and using local plumbing and heating suppliers. Many

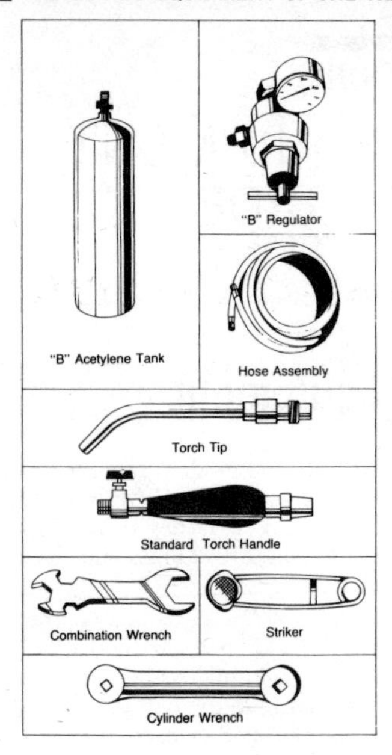

(Figure 1-1) Air-acetylene equipment and tools used in soldering

trade organizations and suppliers offer training seminars at various times during the year. Suppliers and manufacturers usually offer free catalogs illustrating hand tools and materials. These catalogs contain current prices and provide information on the availability of tools and materials.

The tools discussed in this chapter are basic tools used daily by the plumber. There are many more that are not listed here, but the tools described in this section are essential, and they must be used in a professional way to avoid dangerous accidents to the user or fellow workers.

AIR-ACETYLENE EQUIPMENT

Air-acetylene equipment is used to solder copper tubing and fittings and is often referred to as "soldering equipment." Many states still require all hot- and cold-water distribution lines in homes and other buildings to be copper tubing. The plumber, therefore, works with copper tubing often and must learn to use air-acetylene equipment properly to prevent leaks in floors, walls, and ceilings. (See Figure 1–1).

"B" TANKS

"B" tanks contain the gases necessary to solder copper tubing. They contain compressed acetylene gas, which is a colorless gas formed by mixing calcium carbide and water. It has a distinctive, nauseating odor. Each cylinder is equipped with a fuseable plug to relieve any excess pressure (caused, for example, by mechanical pressure or undue heat) and this plug protects the cylinder from exploding. The tanks are also packed with a porous material saturated with acetone, a chemical that absorbs large quantities of acetylene under pressure without changing the nature of the gas.

When acetylene gas is mixed with air and ignited, a hot flame suitable for soldering is produced. With the proper turbo-torch equipment, "B" tanks can also be used for brazing.

A "B" tank holds 40 cubic feet of gas. Another type of tank, called an MC tank or cylinder, holds 10 cubic feet of gas. Adapters allow a plumber to change from one tank to the other.

All acetylene gas tanks should be stored in an upright position. They should never be stored in a closed space. *Improper handling of these tanks can result in explosions.*

COMBINATION WRENCH

Combination wrenches may be used to open or close the "B" tank valve stem. They are used to tighten the regulator and hose assembly. They can also be used to tighten the torch handle and torch tips. Care should be taken not to overtighten the fittings, since brass fittings can be easily stripped.

CYLINDER WRENCH

Cylinder wrenches are also used to open or close the "B" tank valve stem. The valve stem should be opened no more than one-half turn. It can then be shut off quickly in case of emergency. Do not overtighten the valve when closing it, since the stem of the "B" tank can be easily snapped off or broken. Often the cylinder wrench is attached to the tank for safety.

HOSE ASSEMBLY

The hose assembly is tightened (left-hand thread) on the male thread fitting on the regulator. The hose is usually 25 feet long, to allow you to work away from the "B" tank. Hoses should be protected from the flame and kept straight to avoid kinking, or possibly tripping other workers.

REGULATORS FOR A "B" TANK

Regulators for "B" tanks are tightened in a clockwise fashion (standard thread). Pressure regulators are available with fixed pressure or with variable pressure. The pressure gauge on the regulator indicates inside tank pressure, which in some cases is around 250 pounds per square inch (psi). Before the "B" tank cylinder is opened, all tension on the regulator (open torch handle) should be released to prevent a sudden rush of pressure from damaging the gauge or, in extreme cases, even causing the pressure gauge to explode. Care in handling the gauge will prevent the glass, and eventually the pressure indicator, from breaking.

Acetylene gas should never be used at a pressure in excess of 15 psi. It becomes unstable above this pressure and can easily explode on impact. Pressure regulators deliver 10 pounds per square inch of maximum pressure to the torch handle.

STANDARD TORCH HANDLE

Standard torch handles are tightened on the acetylene hose in a counterclockwise direction (left-hand thread). Torch tips are installed into the top of the handle by turning them clockwise (standard thread) until they are hand tight. A round valve stem opens and closes the gas to the tip. Fittings should be tightened to prevent gas leakage, which could result in a fire while you are holding the handle.

STRIKER

Only special strikers should be used to light acetylene torches. Strikers have a removable flint on the tip of the wire. When this flint is pushed across a rough bar on the head of the striker, a spark is created, which ignites the gas. Keep strikers dry and out of mud or dirt at all times.

TORCH TIP

Torch tips mix the acetylene from the tank with the outside air to create a hot flame for soldering. The tips are turned clockwise (standard thread) into the torch handle until they are hand tight. The tips are made from pure drawn copper because copper dissipates heat rapidly.

While soldering, be sure to keep the tip away from dripping flux and solder. If you allow these drippings to enter the tip area, you will have problems with that tip. You will need to clean it or even replace it. To clean the tip, unscrew it (counterclockwise) from the base and clean the screen inside the base.

Torch Tip Sizes. Torch tips come in a variety of sizes. The larger tips allow more gas to be drawn, resulting in more heat.

1. *Very fine tips* are used for small soldering projects on tubing and fittings up to ½ inch.

2. *Fine tips* have a slightly hotter flame and can be used on copper tubing and fittings up to ½ inch.

3. *Medium tips* are commonly used on many soldering projects, including the installation of hot- and cold-water lines. These tips can easily be used on 1-inch tubing and 1-inch copper fittings.

4. *Medium large tips* are used on copper tubing and copper fittings associated with waste pipe. They are also used on large water lines and pressure fittings. These tips can easily be used on tubing and fittings up to 4 inches.

5. *Large tips* are used on large copper tubing and fittings; however, they can be used on 4-inch tubing and fittings, depending on job conditions.

6. *Extra large tips* are also used on large copper tubing and fittings. They are usually used on pipe and fittings in excess of 4 inches.

BARS, STEEL (CROWBARS)

Steel bars are used to remove nails, to apply leverage for removing woods beams or studs, and to straighten pipes. Especially useful for repair, alteration, and renovation work, these tools are usually included in the plumber's toolbox.

(Figure 1-2a) Nail claw

NAIL CLAW

Nail claws are used to remove nails from wood studs. A hammer is used to drive the claws under the nail to be removed. These bars are from 11 to 13 inches long. (See Figure 1–2a.)

(Figure 1-2b) Prybar

PRY BAR

Pry bars are designed for leverage. They are generally used by hand, but a hammer can also be used to force the bar into an opening. Generally, they are 13 to 16 inches long and can be easily used and stored. (See Figure 1–2b.)

WRECKING BAR

Wrecking bars are used for leverage and to lift heavy objects. They come in 24-inch and 30-inch lengths and are usually ¾-inch wide.

BENDERS, COPPER TUBING

Copper tubing benders enable you to bend copper tubing to various angles. This can eliminate the need for a soldered joint, which sometimes is not desirable because of its placement in a floor, wall, ceiling, or other hard-to-reach spot or because a flame cannot be used due to the danger of fire.

There are three basic types of benders: geared-ratchet lever type, lever-type, and spring-type.

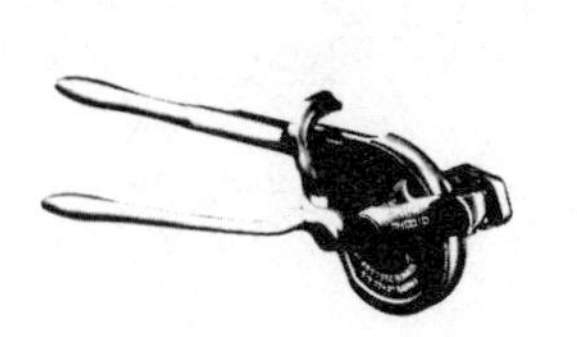

(Figure 1-3a) Geared rachet lever-type copper tubing bender

GEARED-RATCHET LEVER-TYPE BENDERS

Geared-ratchet lever-type benders are used on copper tubing of ⅝-inch, ¾-inch, and ⅞-inch outside diameter (O.D.). To use the bender, place the copper tubing in the grooved wheel and lock it in with the locking bar. Then pull the lever handles apart. Continue until the desired angle is achieved. Then lift the locking bar and remove the tubing. (See Figure 1–3a.)

LEVER-TYPE BENDERS

Lever-type benders are used on smaller tubing, generally ⅜-inch to ½-inch O.D. Place the tubing in the wheel assembly, then pull the handles to create the angle desired. (See Figure 1–3b.)

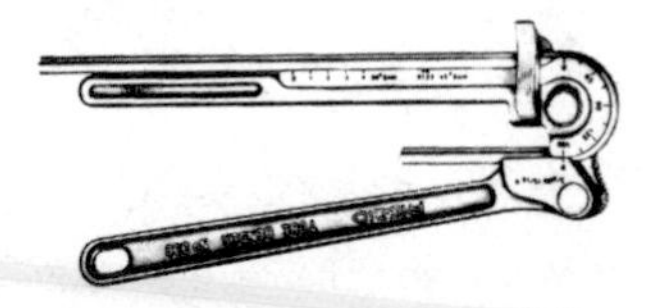

(Figure 1-3b) Lever-type copper tubing bender

SPRING-TYPE BENDERS

Spring-type benders are used mainly on the small, speedy type supply tubes that carry water to fixtures. This small tubing is usually made from copper and brass combinations and is available in copper finish or chrome-plated. The tubing is usually 12 inches long but can be up to 3 feet long. Speedy-type tubing often has a ⅜-inch O.D. and is easily bent. Send the tubing through the inside (by hand) and then bend the spring to the desired angle. The steel spring helps prevent kinking and the collapse of the tubing walls. Spring-type benders are available in sizes ranging from ¼ inch to ⅞ inch. (See Figure 1–3c.)

(Figure 1-3c) Spring-type copper tubing bender

BRACE, STANDARD

A brace allows you to drill holes in wood, concrete, and tile when electricity is not available. It is used with different types of bits. It

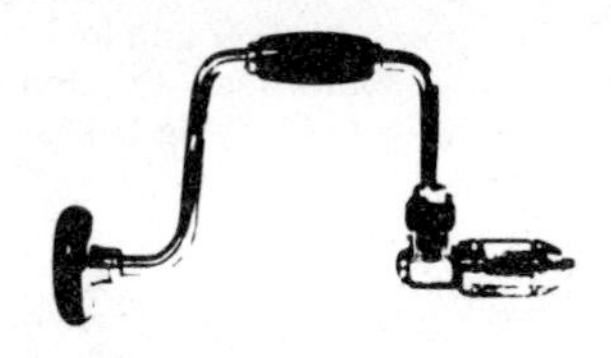

(Figure 1-4) Bit brace

is a handy tool and should be included in the plumber's toolbox. (See Figure 1–4.)

Standard braces are used by slowing rotating the handle in a clockwise manner. When drilling a hole in this way, the bit is drawn in by the worm screw in the head of the bit. To reverse and draw the bit out of the wood, you change the locking device above the chuck to the opposite position. Most locking devices, if left in the middle position, will allow you to drill *and* retrieve. However, you may need to ratchet the bit because of bit size or a lack of room.

BRACE BITS

The bit placed in a standard brace is the part that does the actual drilling. A bit and brace are used to drill holes in wood and to install wood screws. Although electric drills are now commonly used, the manual bit and brace is still a useful tool, especially when, for some reason, electricity is not available.

The bit is placed into the head of the brace and locked in by tightening the chuck. There are several types of bits, each designed for a specific use. Those mentioned below are the ones a plumber most often uses.

AUGER BITS

Auger bits are used to drill holes in wood for small piping projects. These bits come in sizes from ¼ inch to 1 inch. (See Figure 1–5a.)

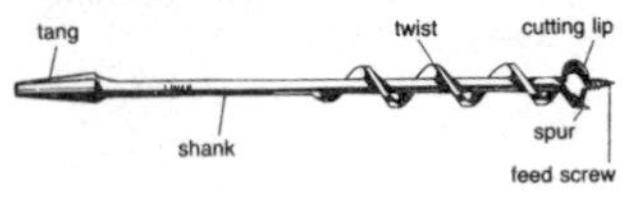

(Figure 1-5a) Auger bit

A sharp auger bit will go through soft wood easily. A dull bit will make drilling the hole more difficult and may result in a rough hole. Be sure to keep all auger bits clean and sharp. The worm screw must also be kept sharp. If the worm screw should break, throw the bit away.

EXPANSIVE BIT

With an expansive bit, the bit size can be increased or decreased by adjusting a screw in the head of the bit. Loosen the screw, adjust the movable bit, and retighten the screw. Expansive bits are used to drill odd-sized holes as well as to drill standard-size holes. When using an expansive bit, be careful to avoid nails and knots in the wood. (See Figure 1–5b.)

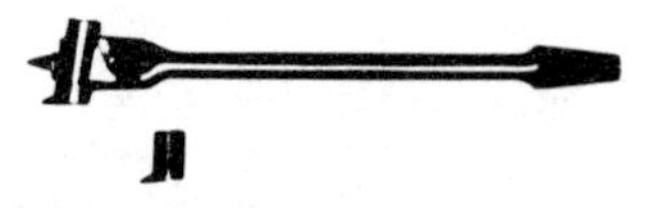

(Figure 1-5b) Expansive bit

SCREWDRIVER BIT

Screwdriver bits are used to install and remove wood screws quickly. They are usually 5 to 6 inches long and come with a standard or Phillips-type tip.

CAST-IRON ASSEMBLY TOOL

A cast-iron assembly tool allows you to join 2- to 4-inch cast-iron soil pipe with neoprene gaskets. This tool is easily used by one person. Care must be taken to clean the tool and to replace worn parts. The parts that usually need replacing are the gripping jaws and the cotter pins. (See Figure 1–6).

To use this tool, place the U-bend or front of the tool past the bell or hub part of the pipe. Place the clamp or stirrup part of the tool on the other pipe. Then push the handle down and force the pipe into the rubber joint.

To remove or disassemble the cast-iron pipe, place the front or U-bend section of the tool against the hub where the rubber gasket is located. Reverse the clamps or stirrups and place them around the same pipe. Then pull the handle down toward the pipe. This should separate the pipe from the gasket.

(Figure 1-6) Cast iron assembly tool

To adjust the clamp for various sizes of pipes, move the pins to the proper places. To adjust the U-bend area, insert the correct U-bend attachment. Move the outside pins to move the clamps and stirrups to the desired length.

CAULKING IRONS

Caulking irons are chisel-like tools used to assemble cast-iron soil pipe by placing and compacting oakum and lead in the hub or bell area of the joint. These irons are designed to be used by hand and with a ball peen hammer. When hitting the head of a caulking iron, be careful to protect the eyes from metal slivers that may fly off. *Safety glasses should be worn at all times when using this tool.* If the head of a caulking iron begins to mushroom, grind it down.

There are several types of caulking irons, each designed for a particular use in the process of making a lead and oakum joint. (See Figure 1–7.)

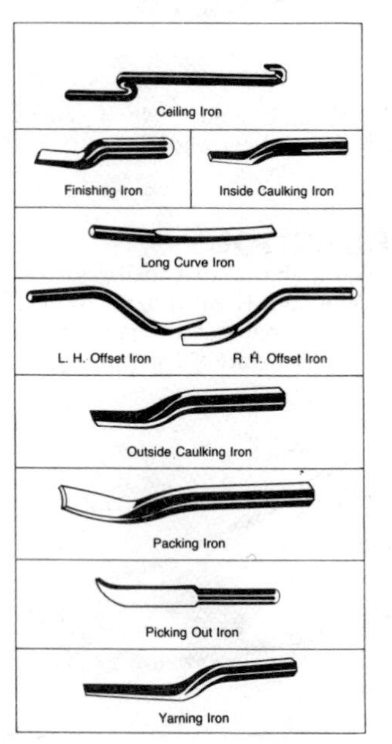

(Figure 1-7) Caulking irons

CEILING IRON

A ceiling iron is used when the oakum needs to be packed tightly and there is no room to use a regular packing iron.

FINISHING IRON

A finishing iron is used to "trim" the poured lead. This iron helps to set the lead and give the finished product a professional appearance.

INSIDE IRON

An inside iron is used to drive or set lead joints and to trim the inside rings near the round pipe.

LONG CURVE IRON

A long curve iron is used to drive the oakum into the hub when there is a problem using regular caulking and yarning irons.

OFFSET IRON

An offset iron allows the plumber to drive oakum into tight places. Available in left-hand and right-hand styles, offset irons are valuable tools, allowing the plumber access to either side and to the rear of the hub to drive oakum in.

OUTSIDE IRON

An outside iron is used to set or drive the lead down and trim it near the outside of the hub.

PACKING IRON

A packing iron is used to drive the oakum down and compact it inside the hub area.

PICKING IRON

A picking iron is used to remove the lead after a bad or cold joint has been poured. It can also be used to take the lead out of a poured lead joint.

YARNING IRON

The yarning iron is the standard iron used to drive oakum into the hub. It is used when there is sufficient room to work.

CHISELS

Chisels serve many purposes. They are useful in making holes and in breaking concrete and cast iron. They are designed to be used with hammers or mauls. When using a chisel, wear safety glasses to prevent metal slivers from damaging the eyes.

The blade of a chisel should always be kept sharp. If the head of a chisel begins to mushroom, grind it down to prevent splintering.

There are many types of chisels. The ones most commonly used by plumbers are discussed briefly below.

BULL-POINT CHISEL

Bull-point chisels are used to break up concrete or to make a hole in a concrete or brick wall. They are designed to be used with a maul.

COLD CHISEL

Cold chisels are used to cut cast-iron soil pipe by hand. They are also designed to be used with a maul. (See Figure 1–8a.)

(Figure 1-8a) Cold chisel

CUTOFF CHISEL

These chisels are used to cut away excess lead after a lead joint is poured. They are designed to be used with a ball peen hammer.

WOOD CHISEL

Wood chisels are used to make holes for pipes in soft wood surfaces. They come in various sizes and lengths and are designed to be used with a regular hammer. (See Figure 1–8b.)

(Figure 1-8b) Wood chisel

Wood chisels, like all chisels, should be kept sharp and clean. An oilstone is usually used for sharpening. A grinding angle of 25 to 30 degrees for rough cuts and 20 degrees for finish cuts is common.

COPPER CLEANING TOOLS

Copper cleaning tools are used to clean the inside and outside of copper tubing and fittings to prepare the material for soldering. In addition to a special tubing cleaner tool, abrasive sandcloth and brushes are used to clean copper piping.

ABRASIVE SANDCLOTH

Abrasive sandcloth is made with artificial abrasives, such as silicon carbide, and natural abrasives, such as emery (a dark, very hard, coarse variety of corundum used for grinding and polishing brass and some other metals). Sandcloth with a fine grit is most often used by plumbers for cleaning copper tubing and fittings. A piece of the sanding cloth is turned around the outside end of the tubing to the depth of the fitting or a little more. This material usually comes in rolls and should be kept dry at all times. (See Figure 1–9a.)

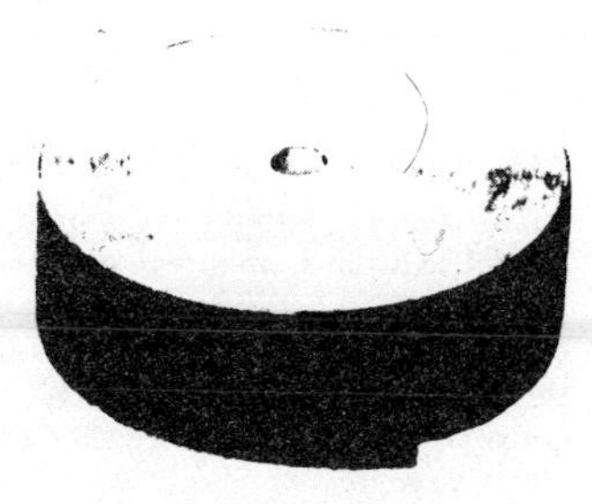

(Figure 1-9a) Abrasive sandcloth for cleaning copper tubing

BRUSHES

Special brushes are designed to be placed inside a copper fitting and turned until the interior of the fitting is clean. Common brush sizes are ½ inch and ¾ inch. The brushes should be cleaned periodically and should be kept dry. (See Figure 1–9b.)

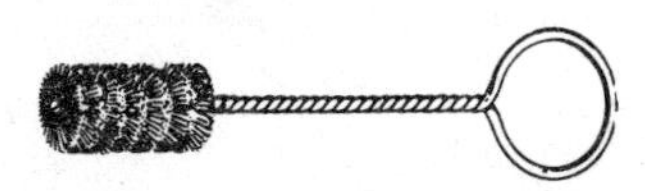

(Figure 1-9b) Brush for cleaning the interior of copper tubing

COPPER-TUBING CLEANING TOOL

This special tool can be used to clean the outside of copper tubing and the inside of copper fittings. The tubing is placed in the hole of the tool and turned

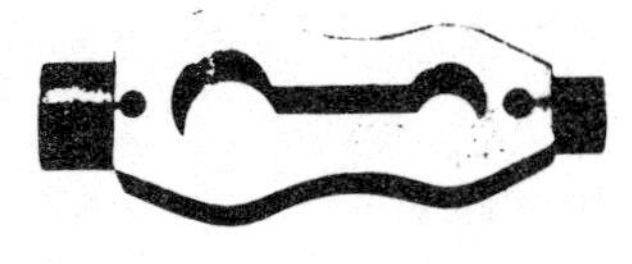

(Figure 1-9c) Copper-tubing cleaning tool

until it is clean. A fitting is placed on the outside of the tool on the round tip and then rotated until it is clean. These tools are used quite often with ½-inch and ¾-inch tubing and fittings. When not in use, they should be kept dry and clean. (See Figure 1–9c.)

DRAIN-CLEANING EQUIPMENT

Drain-cleaning equipment is used to remove or relieve stoppages in house drains, fixtures, or sewers. The simplest of these are plungers. The ball-type and force-cup plungers are placed inside the toilet bowl and pushed and pulled to dislodge any stoppage in the bowl. Most of the other types of drain-cleaning equipment consist of tapes or snakes that are forced through the drain or waterway.

CLOSET AUGERS

Closet augers are a type of snake, and they are placed inside the waterway of the toilet bowl. By rotating the handle in a clockwise fashion and pushing in, you can catch the object that is causing the stoppage, or you may push the object into the main house drain. The auger snake has a very sharp tip, and if you can snare the object causing the stoppage, you can retrieve it by pulling back on the handle. This equipment is used only on toilet-bowl stoppages. (See Figure 1–10a.)

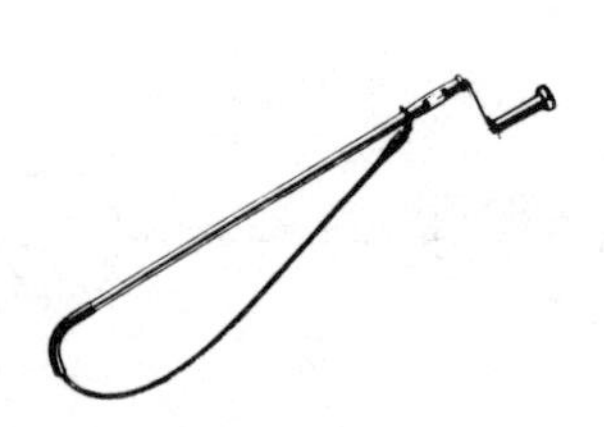

(Figure 1-10a) Closet auger

COIL WIRES

Coil wires are long wires that are inserted into small drain pipes. By rotating the handle in a clockwise manner, you can spiral the wire into and through the stoppage.

ELECTRIC DRAIN CLEANERS

Electric drain cleaners are used to relieve stoppages in kitchen sinks, lavatories, bathtubs, and other fixtures. Feed the snake into the drain opening or waste pipe and tighten the chuck. Then turn on the trigger switch and the snake should spiral into and through the stoppage. This equipment should be used with a 3-wire cord and properly grounded electrical outlet. (See Figure 1–10b.)

(Figure 1-10b) Electric drain cleaner

FLAT SEWER SNAKE

A flat sewer snake is usually placed into an opening in the house drain or house sewer piping and then pushed by hand into the stoppage area. Usually this will break up the stoppage. This flat snake comes in coils of 50 feet or 100 feet, and it does require sufficient room to uncoil it. (See Figure 1–10c.)

(Figure 1-10c) Flat sewer snake

HAND-HELD DRAIN COIL

Hand-held drain coils are fed manually into a small drain. Push the wire into the stoppage and tighten the chuck. Then turn the handle in a clockwise manner and push. The wire should go through the stoppage. (See Figure 1–10d.)

WATER RAM

A water ram uses a charge of compressed air to dislodge a stoppage. By pushing the handle of the ram, you build up an air charge

(Figure 1-10d) Hand held cleaner

inside the tube. You release the air by using a trigger switch. The rush of air should relieve the stoppage, but you must be careful when using a water ram. If there is a severe stoppage in a toilet bowl, the charge of air could pack it in tighter and make relieving the stoppage impossible. Then the entire toilet bowl will have to be replaced.

DRILLS, ELECTRIC

Drills are used to bore holes in wood, masonry, and tile. These holes are necessary for pipe and fitting installations. Power drills are most often used, since they can generally complete the job faster and more accurately than hand drills. Cordless (battery-operated) models have become very popular in recent years. The proper use of this equipment makes work easier and less costly.

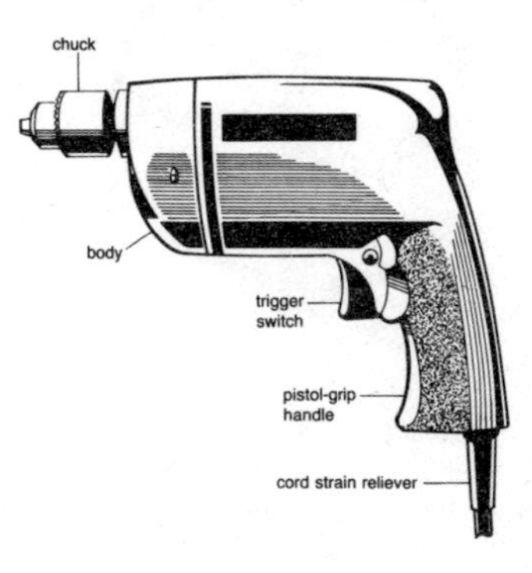

(Figure 1-11) Electric drill

Electric drills come with a variety of features that offer versatility, ease of use, and accuracy. (See Figure 1–11.) They are designated by the maximum drill diameter that the chuck can accommodate. The torque or drilling force increases as the size of the chuck capacity increases; however, the drill speed decreases as the size of the drill chuck increases because of the motor capacity and intended use of the drill. A good, all-purpose drill has a ⅜-inch chuck and variable speeds, and is reversible. The reverse drilling feature allows the bit to be more easily retrieved from the wood and, with special bits, for screws to be withdrawn. Bits are inserted into the chuck and locked into place by a chuck key. (See Figure 1–11.)

The ¼-inch high-speed drills are used to drill into steel, wood, or tile and are used mainly for small holes. When using these drills, be careful not to drill into the copper water lines.

The ½-inch right-angle drills are used for heavy-duty drilling in wood and concrete. The drill head can be adjusted to any angle by loosening the two hex head nuts on the top of the drill. It is then rotated to the proper angle, and the bolts are retightened.

DRILL BITS

A variety of drill bits are available for drilling holes in plastic, wood, tile, and concrete. Bits with worm screws are drawn into and through wood beams and flooring. Plumbers must carry an assortment of those sizes and types of bits that will be used on a daily basis. (See Figure 1–12.)

Bit Extensions. Bit extensions provide added length for drilling in hard-to-reach areas. They are inserted into the chuck and tightened by a chuck key. The bit is placed into the head of the extension, and the hex nuts are tightened by a hex wrench. It's a good idea to carry extra hex nuts because they will eventually need to be replaced.

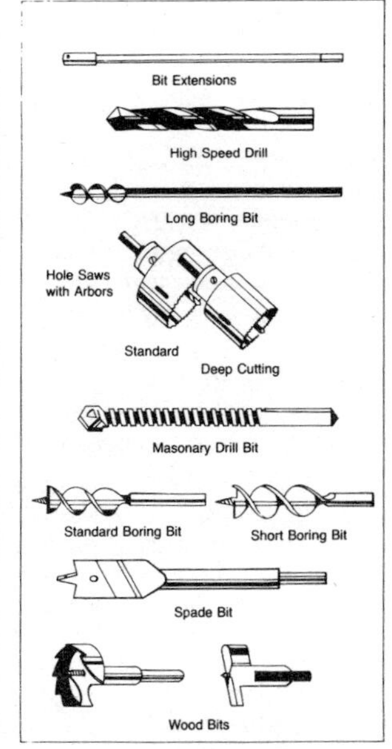

(Figure 1-12) Electric drill bits

High-Speed Drill Bits These bits are installed into the drill chuck of ¼-inch, ⅜-inch, and

½-inch electric drills and are locked with a chuck key. They are used to drill holes in steel, sheet metal, concrete, wood, and tile. When drilling metal, be sure to lubricate the bit to prevent the tip from overheating and burning due to the high speeds. The bits must be sharpened occasionally.

Hole Saws with Arbors. These specialty bits are used to drill large holes, from 1-inch to 3¼-inch diameter, in wood and metal. The arbor allows you to place the bit accurately on the exact spot you wish to drill. These bits are available for all pistol-grip electric drills as well as for the ½-inch right-angle drill. Hole saws work very well and last a long time.

Long-Boring Bits. These bits are used to drill holes through several pieces of wood without using an extension. Sometimes they help to avoid using a ladder. They are inserted into the chuck and tightened with the chuck key.

Masonry Bits. Masonry bits are used to make holes in concrete and tile. They can be used with many different types of drills but are most often used with the ¼-inch high-speed drill. Too high a speed may cause them to burn out. The bit is inserted into the chuck and locked with a chuck key.

Carbide-tipped masonry drills come in various sizes, usually 3/16 inch through ⅞ inch and from 3 to 6 inches long. Wide spiral flutes lift dust smoothly and minimize clogging while the holes are being drilled.

Short-Boring Bits. These bits are used to drill holes in wood or floors and are usually used with the ½-inch drill. They are inserted into the chuck and tightened with a chuck key.

Spade Bits. Spade bits are used to drill very smooth holes in wood. The hole should be started from one side of the wood and continued until half the depth of the hole is completed. Then the bit should be withdrawn and the hole completed by drilling on the other side of the wood. This technique is necessary to make clean and professional-looking holes in wood. It prevents splintering the edges of the wood, top and bottom.

Spade bits are available in sizes from ¼ inch through 1½ inches. Usually used on a ¼-inch drill, the bit is inserted into the chuck and tightened with a chuck key. Try to avoid wood knots and nails when using the spade bit—or any other type of bit.

Standard Boring Bits. These bits are used to drill holes in single-spaced or double-spaced wood areas. They are very handy and can save time. The bit is inserted into the ½-inch right-angle drill and tightened with a chuck key.

Wood Bits. Wood bits are used to drill large holes, up to 3 or 4 inches O.D. They are used in the ½-inch right-angle drill and tightened into the chuck by a chuck key. These are heavy-duty bits, and should be used with care.

DRILLS, HAND

Hand drills are useful when no electricity is available and when light, delicate drilling is needed.

CRANK-OPERATED HAND DRILL

The crank-operated hand drill has a large wheel gear that turns a smaller pinion, which then turns the drill chuck. This hand drill allows you to drill small holes in tile, wood, sheet metal, sheetrock, and masonry when electricity is not available. Where light drilling is required, as for drilling pilot holes in thin, narrow wood trim, the crank-operated drill has an advantage over the power drill in that the weight of the power drill could cause the bit to angle and split the trim. The bit is inserted into the chuck and tightened by hand. The handle is then turned in a clockwise motion to drill a hole. (See Figure 1–13a.)

(Figure 1-13a) Crank-operated hand drill

PUSH DRILL

Push drills are hand-held tools usually used for making small holes in tile, wood, sheet metal, and concrete. They are also referred to as "yankees." Bits of various sizes are usually found in the handle. The bits are inserted into the chuck located at the end of the tool and, on some models, are then locked in. These models have a retractable chuck, which, when pulled back, releases the bit. (See Figure 1–13b.)

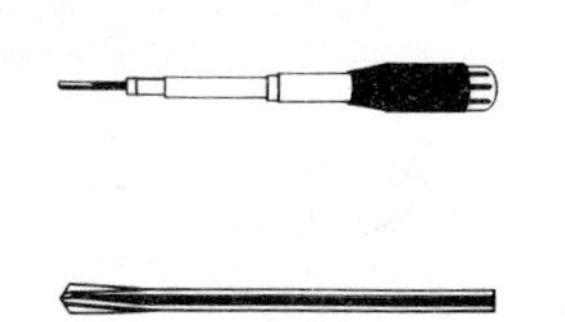

(Figure 1-13b) Push drill and star drill

STAR DRILLS

Star drills are hand-held, chisel-like tools made of steel; they are designed to be used with mauls. The star area rotates when the drill is struck with a maul. Star drills are used to drill in concrete to make round holes for pipe. If the head of the drill begins to mushroom, grind it down to prevent splintering. *Always wear safety glasses when using a star drill.* (See Figure 1–13b.)

DRIVE SOCKETS

Drive sockets are tools used in the installation of fixtures, allowing the plumber to tighten bolts on flanges, wall brackets, hangers, and other pieces of equipment easily and quickly.

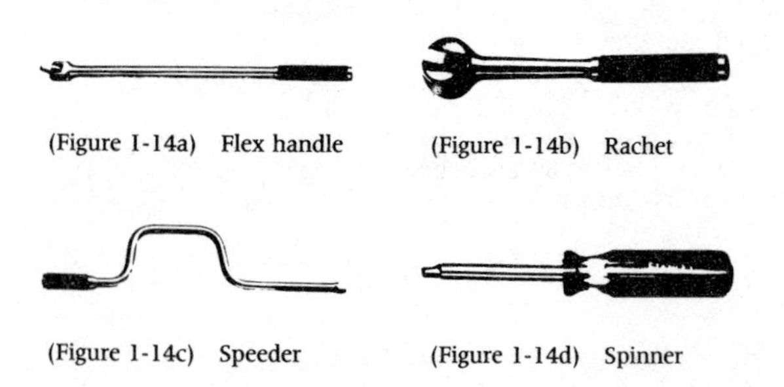

(Figure 1-14a) Flex handle

(Figure 1-14b) Rachet

(Figure 1-14c) Speeder

(Figure 1-14d) Spinner

Different types of drive sockets are available for use in hard-to-reach areas and for use with specific types of bolts. (See Figure 1–14.)

FLEX HANDLE

This type of drive socket is used in tight places where a ratchet cannot be used or when great leverage is required. The socket must be removed from the bolt head and then repositioned as many times as necessary to remove the bolt.

RATCHET-TYPE REVERSIBLE WRENCH

This type of drive socket is used for rapid installation and removal of bolts. The ratchet device can be changed to tighten or loosen bolts by turning it clockwise or counterclockwise. This tool operates in a short arc, or turn.

SPEEDER

A speeder is used to reach bolt heads over a long distance. Turning the speeder clockwise tightens the bolts; turning it counterclockwise loosens bolts. There is no ratchet attachment for this tool, so you need sufficient room to turn the handle.

SPINNER

A spinner is used by hand, much like a screwdriver. It is useful when screwing in flat-headed sheet-metal screws and when working with small objects that require little leverage.

DRIVE SOCKET ACCESSORIES

Drive socket accessories increase the usefulness of the tool, allowing the plumber to complete jobs easily and quickly, many in tight spaces in which standard wrenches are difficult or impossible to use. (See Figure 1–15.) Six-point and twelve-point socket sets are available. Six-point sockets are used for bolts and hex nuts with rounded corners. Twelve-point sockets

(Figure 1-15a) Adapter

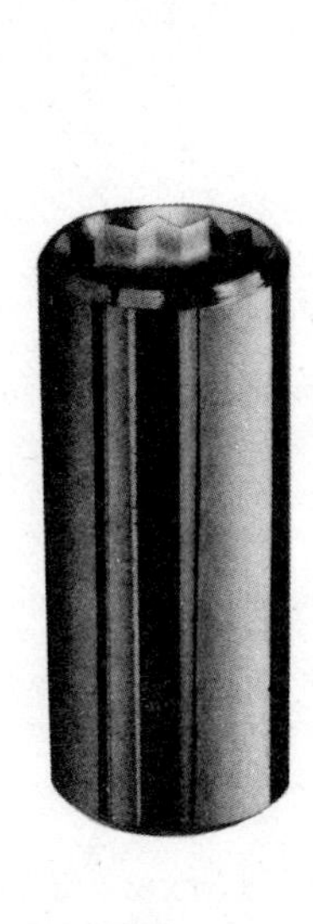

(Figure 1-15b) Deep socket

(Figure 1-15c) Extension

(Figure 1-15d) Universal joint

(Figure 1-15e) 12-point standard socket

are used for hex-nuts and lag screws, among others. They have more turning positions, but care must be taken not to round out the head of a bolt while tightening.

Adapters. Adapters are placed on socket drivers to allow sockets of different sizes to be used.

Deep Sockets. Deep sockets are used when the bolt is longer than a standard-size socket or when the nut is recessed.

Extension Socket. Extensions are used to reach difficult places or to span long distances.

Standard Sockets. Standard sockets are used to loosen or tighten standard nuts and bolts.

Universal Joints. Universal joints are used when flexibility is needed or there are difficult angles to work with.

FILES

Files are used for smoothing the sharp edges of pipe and tubing. They can be used to file inside burrs on steel pipe, copper tubing, and plastic pipe. They can also be used to file down the outside area of a pipe that has a raised area or deep mark on the starting edges created by a dull cutting wheel or saw.

Files are classified by shape, length, and coarseness of the teeth. The coarseness of a file is affected by the length of the file, which is measured from the heel to the point. As the length increases, the tooth size also increases. Common cuts of teeth are single, double, rasp, and curved.

Commonly used file shapes are flat, half-round, round, square, and triangular. The files

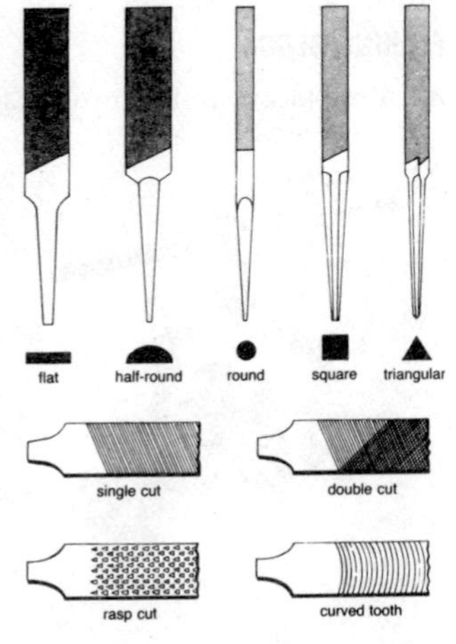

(Figure 1-16) Common types of file

most commonly used by plumbers are the half-round and the round. (See Figure 1–16.)

The half-round files can be placed inside steel pipe, copper tubing, and plastic pipe. By pushing on the file, you can eliminate burrs or raised edges of the pipe. The flat side of the file can be used for smoothing the raised area of the pipe caused by cutter wheels or saws. Round files are used on small pipes and tubing. By inserting and pushing the file, you can easily eliminate inner burrs. Note that files should be used to cut in only one direction.

FLARING TOOLS

Flaring tools are used to make flared ends on copper tubing. Flared fittings make an airtight and watertight connection and are frequently used on copper lines carrying oil and gas. On such lines, soldered joints—which require the use of a flame—cannot be used because of the danger of fire.

On small copper tubing, a flaring block is usually used to make the flared end. On larger tubing, a special flare tool is used.

FLARING BLOCK

A flaring block consists of two metal bars joined together that have holes located along their center. The copper tubing is inserted into the proper-size hole and secured. Then a yoke attached to the bars is turned to make the flared end. The tubing sizes that can be flared in this way are usually ⅛ inch to ½ inch. If the joint is then made properly, there will be no chance of a leak. (See Figure 1–17a.)

(Figure 1-17a) Flaring block

HAMMER-TYPE FLARING TOOL

These flare tools are used on larger copper tubing. The tool is inserted into the center of the tubing, and a maul or hammer used to drive the flare into the tubing. Usually these flare tools are used on water lines, and the flared joint that is made is airtight and watertight. A joint flared in this way is preferable to a soldered joint, especially when the pipes are to be placed underground or under cement. (see Figure 1–17b.)

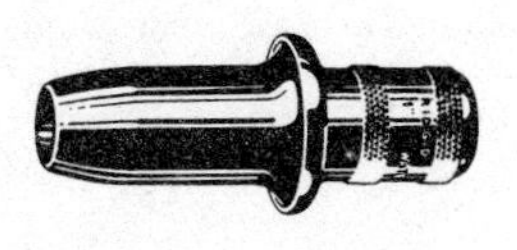

(Figure 1-17b) Hammer-type flaring tool

HAMMERS

Hammers come in a variety of shapes and weights, and each is selected to perform a specific set of tasks. The two most common shapes are the curved claw and straight (ripping) claw. The curved claw is best suited for pulling nails. The straight claw is designed for rough work and can be driven between boards and used to pry them apart.

A hammer head is forged of high-quality steel that is heated to give the poll and face extra hardness. (See Figure 1–18 for hammer parts.) A bell-shaped face, which is slightly convex, minimizes hammer marks on wood when nails are driven flush with the surface.

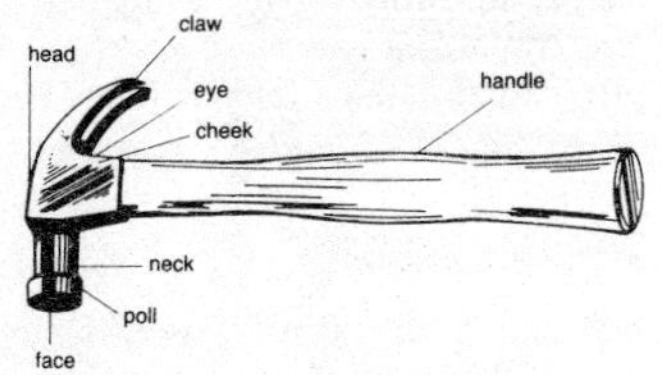

(Figure 1-18) Curved-claw hammer

The size of a claw hammer is determined by the weight of its head. Sizes range from 7 to 20 ounces. The 13-ounce size is popular for general-purpose work.

Certain specialty hammers used by plumbers are described below.

BALL PEEN HAMMER

This hammer is used by plumbers to drive caulking irons when making lead joints. (See Figure 1–19a.)

(Figure 1-19a) Ball peen hammer

HAND-DRILLING HAMMER (MAUL)

This hammer is used when strong striking power is required. It is used with chisels and star drills to cut cast-iron soil pipe and to break or make holes in concrete.

LONG SLEDGE HAMMER

A sledge hammer is useful when breaking up cast iron and concrete. It is swung in a long overhead arc to gain impact power. This tool can be dangerous if you are not careful and accurate in striking the impact area. (See Figure 1–19b.)

(Figure 1-19b) Long handled sledge hammer

LADDERS

Ladders are necessary to reach high places such as beams, attic spaces, and roofs. They are generally made of aluminum or wood and come in a variety of sizes and styles. It's important to use the proper size ladder for the job. *Never work on a ladder higher than the second rung from the top.*

A-FRAME LADDER

A-frame ladders can be used when two persons need to work on a specific project. Each person can use one side of the ladder. Care must be taken to fully open the ladder and to lock the side bars into position. These ladders should be fully opened and used on level surfaces.

EXTENSION LADDERS

Extension ladders are used to gain access to beam areas, lofts, and roofs. The Occupational Safety and Health Administration requires that the distance along the ground to the base of the ladder be at least one fourth of the length of the extension ladder, which forms an angle of approximately 75 degrees. On a windy day, aluminum ladders should be secured by a rope at the top to keep them from blowing down and stranding the workers on the roof. The top of the ladder should be 36 inches above the roof line.

STEP LADDERS

Step ladders are used for easy access to piping areas. Be careful to open the ladder fully and to lock the side bars to prevent collapse of the ladder while working on it. Used only on level ground.

LEAD JOINT EQUIPMENT

Lead joint equipment is used to melt down and pour lead into a cast-iron hub or joint. The use of this equipment is declining, but the plumber engaged in maintenance or repair work must be able to use it safely. (See Figure 1–20.)

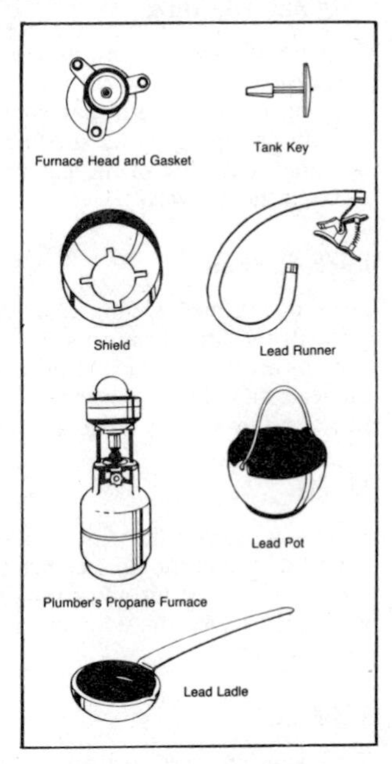

(Figure 1-20) Lead joint equipment

FURNACE HEAD

A furnace head is fitted with a gasket and is screwed onto a propane tank to release the gas from the tank under controlled pressure. A lead pot is placed in a stand or holder located above the furnace head. Care should be taken not to spill lead into the nozzle area. It could extinguish the flame but not shut off the propane gas.

LEAD LADLE

Lead ladles are used to pour hot, molten lead into lead joints. These ladles usually hold up to

5 pounds of molten lead, but they can be bought in different sizes for various projects. They should be dry and warm before inserting them into a pot of hot lead.

LEAD POT

A lead pot first holds the cold lead that is to be melted. Then, once the ignited propane gas has melted the cold lead, it contains the hot lead that will be used for a long period of time. Be careful not to overheat this pot because it can split open and spill hot lead on to the worker or work area.

PROPANE FURNACE

Propane furnaces contain propane gas, which is highly explosive. *Care should be taken to avoid any gas leak,* especially while the furnace is being used. Propane gas valves need constant monitoring to prevent "creeping" while the furnace is running. "Creeping" is the opening or closing of the gas valve due to vibration or pressure through the gas valves.

ROPE RUNNER

Lead rope runners are made of asbestos. Hot molten lead is poured through the top of the rope runner into a horizontal joint. Rope runners are usually oiled to prevent "picking," which occurs when the lead sticks to the asbestos fibers. Removing the lead breaks the fibers and eventually ruins the rope runner.

SHIELD

Shields are steel disks with round openings in the center. They are placed on top of the furnace head to concentrate the heat or flame onto the lead pot for quick melting of the lead ingots, or "pigs of lead." They keep the wind from blowing the flame to one side of the lead pot.

TANK KEY

Tank keys are used to turn the propane gas valve on or off. The key should be left on the valve stem while the furnace is on. That way, in case of a problem, the key is within close reach so that the propane tank can be shut off easily and quickly.

LEVELS

Levels are used to determine the pitch of piping systems. Correct pitch is essential to ensure smooth flow, especially in waste pipes, where waste products must flow quickly and smoothly toward a disposal site. Most waste flows by gravity to a septic tank or sewage-handling facility.

Levels come in a variety of lengths and are made of a number of materials. The body of the level may be made of wood, aluminum, or a special lightweight alloy. Lengths from 3 inches (for a line level) to 6 feet are available. The standard level is 24 inches long. Generally, the longer the level, the greater its accuracy.

A level is used by placing it against the work being checked. The slightly curved vials in the level are filled with nonfreezable liquid, and each vial has a bubble floating in it. The bubble moves within the liquid and rests at the level of the work being checked.

Electronic levels are also available and operate in the same manner as standard levels, except that LED readouts, instead of vials, indicate the level of the surface. Some electronic levels beep when the level or preset degree is attained.

Several types of levels used by plumbers are described below.

CARPENTER'S LEVEL

A carpenter's level is used for long vertical and horizontal piping systems. Common lengths for this level are 24, 36, and 48 inches.

LINE LEVEL

(Figure 1-21a) Line level

A line level is 3 inches long, lightweight, and used on a tightly stretched builder's line. It is used to measure the pitch on a piping system when there are no close marks from which to measure. The line is usually placed on the side of or above the piping system. (See Figure 1–21a.)

PLUMBER'S LEVEL

This special level is more accurate in measuring the pitch of a pipe. It has a movable 1/16-inch measurement bubble and an indicator that measures the pitch or slope of a pipe, usually a drainage line. It also has a standard bubble mounted on the top. Plumber's levels are shorter than 12 inches and are used to make exact measurements of slope on short runs of pipe. (See Figure 1–21b.)

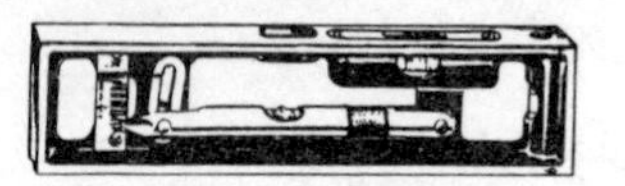

(Figure 1-21b) Plumber's level

TORPEDO LEVEL

The torpedo level has bubbles to indicate the pitch or direction of horizontal and vertical pipes. It is approximately 8 inches long and can be used where other levels won't fit. It has a top view, side view, and a 45-degree vial. It is usually used on short runs of piping. (See Figure 1–21c.)

(Figure 1-21c) Torpedo level

NUT DRIVERS

Nut drivers can be used to install and tighten metal clamps, houdee rim clamps, and sheet-metal (flat-headed) screws. They come in various sizes and lengths and are extremely useful for working in tight places, such as under a kitchen sink. They can be used to quickly install or take apart many types of plumbing equipment. (See Figure 1–22.)

(Figure 1-22) Nut driver

OILERS

Oilers are containers that hold the oil used to lubricate and protect the dies and threads of a pipe during the threading of steel pipe.

The oil used for this operation is usually sulfur-based and is designated as "cutting oil."

OIL CAN

To use an oil can, push your thumb against the bottom of the can and aim the spout into the die areas. The excess cutting oil falls directly on the surface below the die areas, so be sure to clean the floor area to prevent injury from someone slipping and falling.

PAN OILER

This type of oiler captures and filters cutting oil so that it can be reused. Load the oil into the bottom pan, which has a tube attached to allow the oil to flow to a levered handle. Press the lever on the handle, and aim the tip into the threading area. The pan is placed beneath the threading area so that any excess oil falls into it. The oil then passes through a filtering mechanism at the top of the pan. It then can be reused.

UTILITY OILER

This is a compact gallon container that is easily portable. The container has a tube with a levered handle attached to the end so that the oil can be directed to the work area.

PIPE CUTTERS, STEEL

Steel-pipe cutters are hand tools for cutting steel and brass pipes. Sharp-edged wheels are located within this clamplike device and are forced inward by screw pressure. (see Figure 1–23a.)

(Figure 1-23a) Heavy-duty pipe cutter

Regular steel-pipe cutters have one cutting wheel. The cutter is placed on the steel pipe and gently tightened onto the pipe surface. Then rotate the cutter 360 degrees. Gently tighten the handle until the pipe is cut. These cutters are usually used to cut pipe ⅛ inch to 2 inches. Larger cutters can be used to cut larger pipes. Cutters are very durable, and usually only the wheel has to be replaced if it breaks or becomes dull.

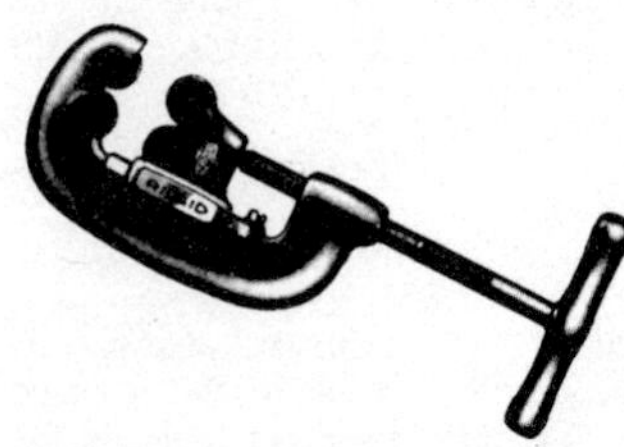

(Figure 1-23b) 4-wheel pipe cutter

FOUR-WHEEL PIPE CUTTER

This cutter is used the same as other cutters except that you need to rotate the cutter only 90 degrees to fully cut the pipe. These cutters are used mainly in alteration work or in tight places. (See Figure 1–23b.)

(Figure 1-23c) Wide roll pipe cutter

WIDE-ROLL PIPE CUTTER

This cutter is used on electric pipe machines to prevent "spiraling" on the pipe while it is turning. Once the cutter is placed on the pipe, you tighten the handle and rotate the cutter 360 degrees. Gently tighten, and proceed in this manner until the pipe is cut. (See Figure 1–23c.)

PIPE CUTTERS, CAST-IRON

Cast-iron soil pipe can be cut using a hammer and chisel. You mark the area to be cut and then hammer a sharp chisel into the cast iron until the pipe breaks. After each stroke, you rotate the pipe until a complete circle is made. Continue this process until the cast iron separates on the mark. Gone are the days, however, when you cut cast iron only by hand. Electrical and mechanical tools help the plumber to complete the job faster and easier.

CHAIN-TYPE CUTTER

This cutter has a chain with small cutting wheels mounted within each link. You wrap the chain around the pipe and gradually ratchet the chain so that there is equal pressure on all sides of the pipe. The shearing force finally causes the pipe to separate. (See Figure 1–24.)

(Figure 1-24) Chain-type cutter

HYDRAULIC CUTTER

This cutter is easy to use. You place the chain around the pipe and secure it. Then you pump the foot pedal, gradually increasing the pressure. Eventually the pipe will snap.

SQUEEZE SCISSOR-TYPE CUTTER

To use this cutter you will need a lot of room on either side of the pipe. Wrap the chain around the pipe and secure it. Then use leverage to squeeze the handles together. Using this cutter requires a great deal of strength. Be careful not to lose your balance while using this equipment; you could very easily strain your muscles or hurt your back.

PIPE EXTRACTOR

Pipe extractors are used to remove pipe where there is not enough room to use a wrench on the pipe. For example, if pipes are placed inside walls and the walls are then completed flush with the pipe nipple there is no room for a regular pipe wrench to grip the pipe for removal. (See Figure 1–25.)

(Figure 1-25) Pipe extractors

To use a pipe extractor, place the extractor inside the pipe nipple. Then tap the extractor with a hammer to make sure the extractor won't turn inside the nipple. Next, take a wrench and place it on the square sides of the extractor and turn counterclockwise to remove the pipe from the fitting. Sometimes this method does not work because the pipe has rusted in place. When this happens, walls, floors, or ceilings have to be broken to remove the pipes.

PIPE TAP

Pipe taps are used to clean and straighten the threads on the inside of brass and steel pipe fittings. If these fittings have been lying around a while or have been dropped, the inside threads sometimes become rusted or damaged. This can cause leaks, so it is important to align and clean the threads properly. (See Figure 1–26.) Insert the pipe tap into the fitting. Rotate the tap clockwise by hand or with a wrench. Pipe taps come in standard pipe sizes, ⅛ inch to 2 inches inside diameter (I.D.). Pipe taps should be cleaned with a steel brush.

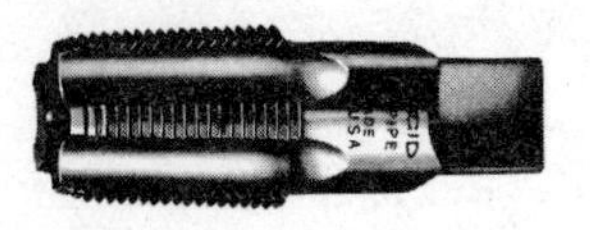

(Figure 1-26) Pipe tap

PIPE THREADERS, MANUAL

Pipe threaders are used to put a standard tapered thread on the ends of steel and brass pipes. They can be used by hand or in conjunction with electric pipe vises. Electric threaders are also available. Various types of manual pipe threaders are described below.

(Figure 1-27a) Enclosed ratchet threader

ENCLOSED RATCHET THREADER

This type of threader is used with interchangeable, variously sized die heads. The die heads come in standard pipe sizes from ⅛ inch to 1 inch. They drop in and lock in place on the end of a handle. This threader is handy for use in tight places. (See Figure 1–27a.)

EXPOSED RATCHET THREADER

This threader is used in the same way as the enclosed ratchet threader but requires a little more room for use because of the die heads. Sizes run from ⅛ inch to 1 inch. The dies are fixed and cannot be moved. (See Figure 1-27b.)

(Figure 1-27b) Exposed ratchet threader

QUICK-OPENING THREADER

The quick-opening die set is used with the electric vise. There is a cam-type lever on top of the equipment. To adjust the die set, loosen the lever, pull the handle to the desired die setting, and then retighten the lever. Next, move the guide set on the back to the proper pipe size. Close the die set (top handle) and slide the guide onto the pipe. This equipment usually comes in three different sizes. One quick-opening die set can be used for ⅛-inch to ½-inch pipe; another can be used for ½-inch to 1-inch pipe; and a third can be used for 1-inch to 2-inch pipe. (See Figure 1–27c.)

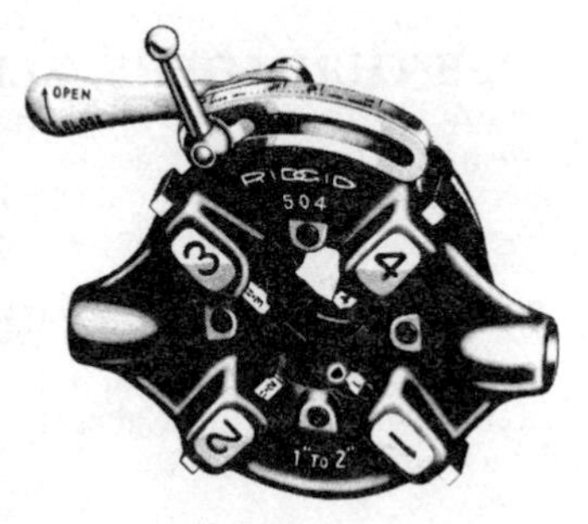

(Figure 1-27c) Quick opening threader

RATCHET-TYPE LARGE-PIPE THREADER

This threader is used on pipes from 1 inch to 2 inches in size. Make adjustments by rotating the guide section, lining up the hole with the indicator of pipe size, and rotating the guide back to the standard setting. Then guide the equipment onto the pipe, close the die-set thumb lever, tighten the butterfly screw onto the pipe, and ratchet in a downward, clockwise manner.

(Figure 1-27d) 3-way threader

THREE-WAY PIPE THREADER

These dies are preset and locked into place. Slide the pipe into the proper hole, and the threader lines up the pipe with the proper threads. Two types are available: one for ⅛-inch to ½-inch pipe and one for ½-inch to 1-inch pipe. (See Figure 1–27d.)

PIPE THREADERS, ELECTRIC

Electric pipe threaders are used to cut threads quickly and efficiently when a large quantity of pipe must be threaded. Because these are electrical devices, care should be taken to prevent electrical shock. Use a three-prong safety plug and grounded outlet.

PORTABLE PIPE-THREADING MACHINE

This machine has all types of threading and cutting equipment on it. It is easily moved and can be transported on a truck. It may also be used as a vise.

WHEELED PIPE THREADER

A wheeled pipe threader is used on large construction sites or in shops. All of the threading and cutting equipment is mounted directly on it, and it can be wheeled to various locations on the job site. Also mounted on this type of threader are reamers and oilers. Wheeled pipe threaders are expensive, but they are well worth the investment when used for large projects or over a long period of time. When used properly, these threaders are efficient, labor-saving tools.

PIPE VISES

Pipe vises are used to hold pipe tightly in place during the threading operation. They are mounted on a stand, usually a tripod-type, and are easily carried and stored. There are two types: chain vises and yoke vises.

CHAIN VISES

Chain vises are used for many purposes. They open fully and can readily be used for all sizes of pipe, copper tubing, and plastic pipe. They may even be used for 4-inch and 5-inch pipe.

YOKE VISES

Yoke vises are also used to secure pipe and tubing, but 2-inch I.D. pipe is the largest size that you can insert into the jaws of this vise. The advantage of this vise is that once the pipe is tightened, it does not have a tendency to turn while it is being threaded.

PLASTIC-PIPE TOOLS

Tools are available that are specifically designed to cut and deburr plastic pipe and tubing. (See Figure 1–28.) The inside of plastic pipe must be smooth so as not to restrict the flow of fluids going through it. Sharp edges could snag materials flowing through the pipe and create a stoppage.

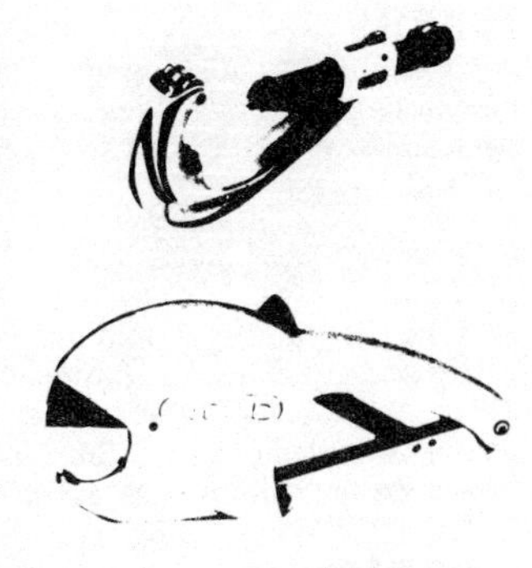

(Figure 1-28a) Tools for cutting plastic pipe

CUTTERS

Standard cutters for plastic tubing are placed around the pipe and rotated 360 degrees. To tighten this type of tool, gently turn the handle clockwise. Plier-type cutters are also available, and these are used like scissors. You exert pressure by closing the two handles. These tools are designed to cut tubing from ⅛ inch to 1½ inch O.D. (See Figure 1–28a.)

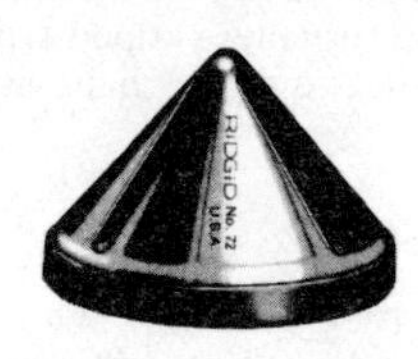

(Figure 1-28b) Deburring tool

DEBURRING TOOLS

Deburring tools are used to remove the inside burr, or raised edge, caused by cutting the pipe. Insert the tool into the center of the pipe and twist it clockwise until you have a smooth end. These tools are used to deburr all sizes of plastic tubing up to 2-inch I.D. (See Figure 1–28b.)

PLIERS

Pliers are used to hold, tighten, clamp, and loosen many types of pipe, fittings, and miscellaneous items. They come in a variety of sizes and shapes; the right tool depends on the nature of the job at hand.

LOCKING PLIERS (VISE GRIP)

Locking pliers grab and hold the work when a lever on the handle is engaged. To adjust the jaw size, merely turn the handle screw. This tool is especially useful when a strong grip is needed to hold an object for nailing or welding. There are straight-jaw and curved-jaw models in various sizes, as well as a host of specialty vise grips.

(Figure 1-29) Slip joint pliers

SLIP-JOINT PLIERS

Slip-joint pliers are so named because of the two-position pivot construction that provides normal and wide jaw openings. They are used for general gripping work, especially for small projects. (See Figure 1–29.)

WATER-PUMP PLIERS

Water-pump pliers are used to tighten and loosen many different sizes of lock nuts. They have a multiposition pivot that permits the jaws to be adjusted for objects up to 2 inches. Often they are used in place of a pipe wrench.

PLUMB BOB

A plumb bob is a metal weight suspended from a string or a line. The weight pulls the line into a true vertical position for laying out hole sites for vertical piping and tubing and for centering fittings. The point of the plumb bob must always be directly below the point from which it hangs. When the plumb bob is first suspended, it will rotate; when it stops moving, it is ready for checking and marking the hole site.

REAMING TOOLS

Reaming tools are used to remove the inside burr made when pipe or copper tubing has been cut. The burr could snag or catch material flowing through and cause a stoppage. Smooth interior pipe walls prevent flow restriction. (See Figure 1–30.)

(Figure 1-30) Reaming tools

BRASS AND STEEL REAMERS

Brass and steel-pipe reamers usually have handles. The reamer is inserted into the pipe, and the handle is rotated in a clockwise direction. During the reaming, gradual pressure is exerted until the burr is smooth with the interior wall of the pipe. Standard reamers can ream ⅛-inch to 2-inch I.D. pipe sizes. Larger reamers can be used on larger pipe sizes.

COPPER REAMING TOOLS

Reaming tools for copper tubing are cylindrical. The tool is inserted into the piping and rotated clockwise to reduce the inside burr.

The tool can also be placed over the outside of the pipe and rotated to grind the outside area to make it smooth and able to enter the fitting easily.

RULERS AND TAPES

Rulers are used to measure and mark pipe for cutting and for estimating work and making plans. No project can be completed successfully if measurements are not accurate. To ensure accuracy, the rule or tape must be read correctly, and the rule must be kept clean for easy reading.

ARCHITECT'S RULE

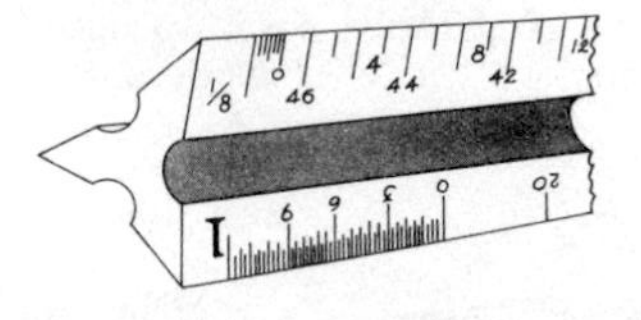

(Figure 1-31a) Architect's rule

The architect's rule is used to draw the plans for blueprints. It is also used to take measurements from plans when you need to use the same scale as used in the drawings. A ⅛ scale is visible from the top edge. This means that ⅛ inch is equal to 1 foot if not otherwise noted on the drawing. (See Figure 1–31a.)

The units are numbered in intervals of 4; each long tick mark represents 1 foot. So a reading between 0 and 23 gives 23 feet. Reading backwards from 0 on the graduated scale, you find that a long tick mark on the graduated scale halfway between 0 and the end of the scale. That halfway mark represents 6 inches or ½ foot. Each of the smaller lines in the graduated portion from 0 to ⅛ represents 2″. (See Figure 1–31b.)

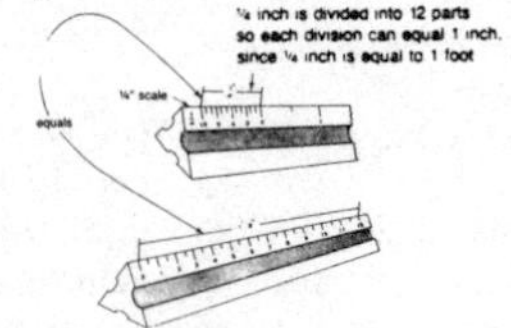

(Figure 1-31b) Using an architect's rule

EXTENSION RULE

This rule is essential and is carried at all times while on a job. It is a folding wood rule and is normally 6 feet long when extended fully. It is equipped with a 6-inch metal extension slide that is useful in taking inside measurements. Folding rules are used for general measuring purposes and are especially handy where rigidity is needed, such as in measuring long distances.

STEEL-TAPE RULE

The steel-tape rule usually ranges in length from 6 to 25 feet, but lengths of 50 to 100 feet are also available. Tape blades come in various widths, but the 1-inch blade is generally preferred because of its stiffness.

Because the tape is flexible, it can be used to measure round as well as straight objects. It is compact, can be clipped to work clothes, and is easily carried on the job.

SAFETY EQUIPMENT

Safety equipment is used on the job site to protect the worker from harm. Certain equipment is required by insurance companies and by state and federal laws. Safety equipment used by plumbers includes:

- steel-tipped boots to protect the toes and feet from heavy falling objects
- face shields to protect the face and eyes while grinding steel or soldering
- glasses with safety lenses to protect the eyes while soldering, brazing, and cutting pipe and tubing
- gloves to protect the hands when handling pipe and fittings
- goggles that can be worn over eye glasses or contact lenses
- hard hats, which are required on most job sites, to protect the head from falling objects

SAWS, HAND

Saws are used to cut wood, metal, copper tubing, and plastic pipe. Hand saws of all types are used by the plumber almost daily on the job. The saw should be kept clean and the blade sharp. Saw blades should be replaced periodically.

COMPASS SAW

Compass saws have narrow blades for cutting circles in plasterboard and wood. The typical compass saw has 12- to 14-inch blades with 8 to 10 teeth per inch. Because compass saws do not have distinct frames to hold blades, they are not limited to working near the edge of a panel. Thus, they can be used to cut openings in floors or walls for pipes or electrical outlets, with the cut starting from a bored hole. A vertical stroke is used to begin the cut. As the cut progresses, the saw is brought to about a 45-degree angle. If you start a cut from the edge, the saw can be at that angle from the beginning. (See Figure 1–32a.)

(Figure 1-32a) Compass saw

CROSSCUT SAW

Every plumber needs a good crosscut saw with a tooth size of 8 to 11 points. (Tooth sizes in saws are given as points or teeth per inch.) Crosscut saws, as the name implies, are designed to cut across the wood grain. Their teeth are pointed and have the same effect as a knife cutting into the wood fibers. In high-quality saws, the teeth are precision-ground to tiny points. In low-quality saws, the teeth have the same shape as the better-grade saws but are not precision-ground for clear cutting. A saw with a low point number cuts fast but leaves a rough surface. A saw with a high point number works more slowly but leaves a smoother surface. To reduce sawing friction and boost efficiency, teeth are alternately bent outward about a quarter of the blade thickness to opposite sides. This produces a cut that is slightly wider than the blade thickness and lets the saw cut freely. (See Figure 1–32b.)

(Figure 1-32b) Crosscut saw

DRYWALL OR SHEETROCK SAW

A drywall saw has large, specially designed teeth for cutting through paper facings, backings, and the gypsum core of drywall. It also has rounded gullets (openings in the blade) to prevent clogging from the gypsum material. A drywall saw is used by plumbers to open holes in walls for waste and water pipes.

First, mark the line to be cut on the wall. Then, place the point of the saw blade where you want the hole to be cut. Slowly work the blade up and down or left to right, depending on the direction you are going to cut. Apply more pressure as you force the blade tip through the drywall. However, don't apply too much pressure since this might break the drywall or cause damage behind it. After the blade tip is through the drywall, start the sawing action along the marked line. (See Figure 1–32c.)

(Figure 1-32c) Drywall saw

HACKSAW

Hacksaws are used to cut steel pipe, copper tubing, sheet metal, and nails, bolts, and other metal fasteners, as well as exterior and interior trim. Most hacksaws have adjustable frames to hold several sizes of blades.

Hacksaw blades are made of high-speed steel, tungsten alloy steel, molybdenum steel, and other special alloys. Generally, blades with coarse teeth are used on thick metal, and blades with finer teeth are used on thinner metal. At least two teeth should always be in contact with the material to prevent a thin section of metal from hooking between the teeth and breaking them. When cutting very thin stock, tip the saw so that the teeth are in contact with a part of the surface, rather than the edge. (See Figure 1–32d.)

(Figure 1-32d) Hacksaw

As a rule, you should use a blade with 14 teeth per inch for brass, aluminum, cast iron, and soft iron. A blade with 18 teeth per inch is recommended for mild steel, tool steel, and general work.

Like wood-cutting saw blades, hacksaw blades have a set or angle to the direction of the teeth. This provides clearance for the blade to slide through the cut. Blades should be installed with teeth pointing forward, away from the handle.

JAB SAW

Jab saws are used for cutting brass toilet bolts. They are especially good for hard-to-reach cutting jobs. They use standard hacksaw blades (even broken ones that can no longer fit into a standard hacksaw).

UNIVERSAL SAW

Universal saws are used to cut plastic pipe and tubing. They are also used for cutting laminates, veneers, and plywood. The rounded serrated tip of the universal saw blade also allows you to use it for reaming the burr from the inside of plastic pipe.

SAWS, POWER

Power saws are used for many purposes—to cut wood as well as various types of pipe and tubing. Heavy-duty portable band saws and heavy-duty circular saws and jig saws are used for large jobs. But the power saw most frequently used by the plumber on a regular basis is the electrical reciprocating saw.

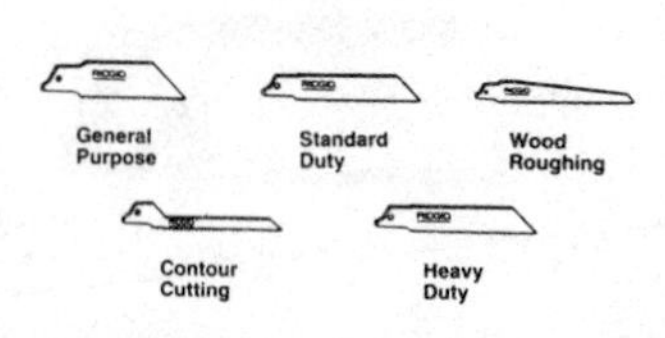

(Figure 1-33) Reciprocating saw and blades

ELECTRICAL RECIPROCATING SAW

The electrical reciprocating saw is useful in cutting all types of materials. It cuts wood, pipe, and tubing quickly and easily. Saw blades come in all sizes. Select the proper blade and insert it in the front of the tool, locking it in with a hex wrench.

Be sure the blade is sharp; dull blades make the job harder and can damage the saw. Some saws have a switch on the handle for two-speed operation. (See Figure 1–33.)

The reciprocating saw is often used to make a hole in wood flooring or in a roof. To do this, place the saw blade at an angle, approximately 15 to 20 degrees. Start the blade through the wood. Once the wood has been penetrated, raise the angle of the saw to a vertical position to finish the cut. The shoe of the saw must be flush with the wood you are cutting to protect yourself from the jabbing action of the saw arm. It also allows the blade to move freely into the wood without binding. If it binds, the blade will pull out of the blade holder.

SCREWDRIVERS

Screwdrivers are used to tighten and loosen all types of screws and bolts. They come in a variety of sizes and tips and are used almost daily. Sizes are specified by the length of the blade, measuring from the tip to the ferrule (the metal band around the wooden handle at the point where the blade enters the handle) or (if the handle is manufactured from a synthetic material) to the point where the blade enters the handle. The 3-, 4-, 6-, and 8-inch screwdrivers are the most commonly used.

Phillips screwdrivers are used to tighten or loosen by hand all types of screws and bolts with Phillips heads. The size of a Phillips screwdriver is given as a point number ranging from 0, the smallest, to 4. Both Phillips and standard screwdrivers come in short ("stubby") lengths for use in tight or hard-to-reach areas.

The slot of the screw and the blade of the screwdriver must be matched carefully. The width of the screwdriver tip should be equal to the length of the bottom of the screw slot.

SPECIAL TYPES OF SCREWDRIVERS

In addition to standard and Phillips screwdrivers in various sizes, there are several other types of screwdrivers, designed for particular purposes. Some are commonly used by plumbers.

Offset Screwdriver. This tool provides more leverage than straight-handled screwdrivers and can be used in tight or hard-to-reach places.

Screw-Holding Screwdriver. A screw-holding screwdriver is handy when working in hard-to-reach places. The screw is held in place by a clip or magnet while the screw is started in the hole.

Spiral Ratchet Screwdriver. A spiral-ratchet screwdriver can drive a number of screws rapidly. You simply push down on the handle and the spiral ratchet spins the blade. These screwdrivers come with various bits that can be mounted in the chuck to work with different screw sizes and types.

A spiral ratchet screwdriver can be set in reverse to remove screws or locked and used as an ordinary screwdriver.

SNIPS

Snips are used to cut metal, flat copper, lead, and aluminum flashings. They should not be used for cutting anything they were not made to cut, and they should never be used to cut wire. Using them to cut the wrong materials will cause the cutting edge to dull or break, making the tool useless. The two basic types of snips are aviation snips and straight snips.

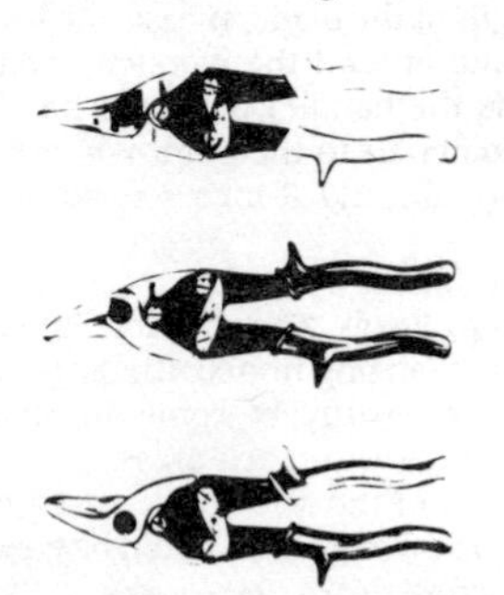

(Figure 1-34a) Aviation snips

AVIATION SNIPS

Aviation snips have a narrow nose and come in different cuts or curls. The handles are color coded to indicate the direction of the cut: right-hand snips have red handles, left-hand snips have green handles, and straight snips have yellow handles. All types of aviation snips are operated by opening and closing the handles. (See Figure 1–34a.)

STRAIGHT SNIPS

Straight snips have more of a scissorlike appearance and are operated by opening and closing the two handles. They are heavy-duty tools and are most often used on metals that are thicker than those cut by the aviation snips. (See Figure 1–34b.)

SQUARES

Squares are used as straightedges for marking and aligning angles of pipe and fittings. They are also used to mark wall backings for fixtures.

(Figure 1-34b) All-purpose straight snips

COMBINATION SQUARE

Combination squares come in a variety of sizes and are used to check the squareness of surfaces and edges. The adjustable sliding blade allows this square to be conveniently used as a gauging tool. The measurements and angles are clearly marked and easy to read. You can measure and cut at precise angles of 45 and 90 degrees. These squares are also helpful in marking studs so that they can be cut square for installation in the walls or for their use in wood backing. (See Figure 1–35.)

(Figure 1-35) Combination square

STEEL SQUARE

Steel squares are used to line up and mark flanges. They are also useful in marking angles for offset pipe and fitting installations. Two steel squares are used facing each other to align pipe during fitting. (See Figure 1–36.)

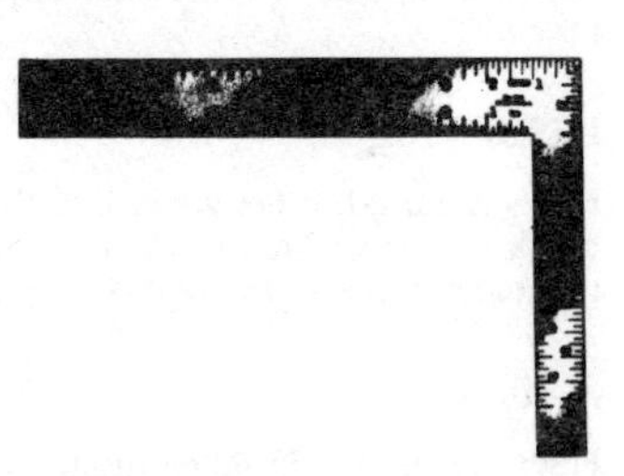

(Figure 1-36) T-square

SWAGING TOOL

Swaging tools allow the plumber to enlarge the inside of copper and brass pipe or tubing. They are generally used to make a coupling on the end of the pipe or tubing. Swaging tools come in various sizes, but they are usually used to enlarge ½-inch and ¾-inch copper tubing and 1½-inch rough brass tubing. (See Figure 1–37.)

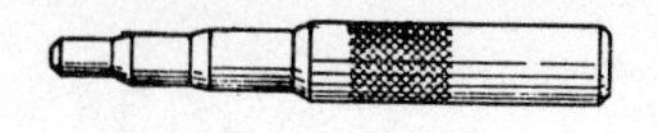

(Figure 1-37) Swaging tool

To enlarge pipe or tubing, insert the swaging tool into the center of the pipe or tubing. Using a hammer, hit the end of the tool until it enlarges the inside diameter of the pipe the desired amount.

TESTING EQUIPMENT

Testing equipment is used on the various pipes and materials of a plumbing system. The testing of piping systems is required by code. Plumbing and sanitary inspectors insist on testing procedures to protect public health.

AIR PUMP (HAND-HELD)

This pump is used manually to pump air into the piping system to check for leaks, cracks, or bad solder joints. Normal test pressure is 50 psi.

CAPS (I.P.S.) (IRON PIPE SIZE)

Caps are used to seal the ends of pipes so that the system can be tested.

HOSES

Hoses are used to get water into the drainage or copper piping to check or test the system. Water tests require filling the system to the extreme top of the highest vent pipe.

PLUGS (I.P.S.)

Plugs are usually installed inside fittings to allow the system to be tested.

PRESSURE GAUGE

This gauge is used to indicate how much water or air pressure is on the system. Usually, 50 pounds of pressure is required for inspection.

TEST BALLS

Test balls are used to hold water pressure on the drainage system. Air is usually pumped in to expand and hold them in place.

TESTING PLUG

A testing plug is inserted into the soil pipe, and the wing nut at the stem of the plug is tightened. This expands the rubber ring and makes an airtight and watertight joint necessary for testing the system. Test plugs come in many sizes and effectively hold the water in the drainage system for testing.

TUBING CUTTERS

Tubing cutters are used to cut copper tubing. They come in a variety of sizes and require routine maintenance. Replacing broken or dull wheels and oiling the moving parts help to make cutting copper tubing easier.

INTERNAL TUBING CUTTER

This cutter is inserted into the middle or inside diameter (I.D.) of the pipe or tubing. The handle of the cutter is tightened gently and then the cutter is rotated clockwise to cut the pipe from the inside to the outside of the pipe walls. (See Figure 1–38a.)

LARGE TUBING CUTTER

This cutter usually cuts tubing from 1½ to 4 inches in size. It is placed over the outside of the tubing, and the handle is tightened gently while the cutter is rotated 360 degrees around the outside of the pipe. (See Figure 1–38b.)

(Figure 1-38a) Internal tubing cutter

(Figure 1-38b) Large tubing cutter

(Figure 1-38c) Midget tubing cutter

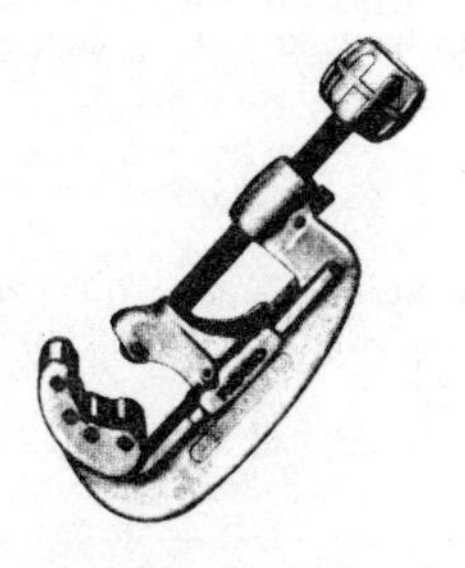

(Figure 1-38d) Quick-acting tubing cutter

(Figure 1-38e) Regular tubing cutter

MIDGET TUBING CUTTER

This cutter is used on copper tubing ranging from ¼ inch to ¾ inch in size. It is placed over the outside walls of the copper and then rotated 360 degrees to cut the tubing. This cutter is handy for working in tight spaces. (See Figure 1–38c.)

QUICK-ACTING TUBING CUTTER

This type of cutter has a release latch on the handle, which when depressed allows you to rapidly adjust the wheels and rollers to the proper tubing size. The tool is placed on the outside of the tubing and tightened gently. A complete turn of 360 degrees will finally cut the tubing. This cutter comes in a variety of sizes for use on copper tubing. (See Figure 1–38d.)

REGULAR TUBING CUTTER

This cutter is placed on tubing ½ to 1 inch in size and is rotated 360 degrees. It generally has a reaming tool (for reaming the inside of the copper tubing) and a hole (for the cylinder key) in the reaming blade to turn the "B" tank on and off. (See Figure 1–38e.)

WRENCHES, GENERAL PURPOSE

Wrenches are used to grip nuts and bolt heads that must be held for fastening and unfastening. They are available in a variety of forms designed to grip everything from pipes to spark plugs or to work in hard-to-reach areas.

When selecting a wrench for a particular job, make sure it is the correct size. A wrench may appear to be the proper size because it is wedged on the points of the nut. However, when pressure is applied to a wrench that is the wrong size, the wrench will slip and ruin the nut.

The wrenches described in this section are used continually by the plumber for a variety of jobs. They are useful in maintenance, repair, and installation work. They need to be kept clean and should be oiled occasionally. This keeps them functioning properly so that the plumber can complete the job quickly and efficiently. Figure 1–39 shows different types of general-purpose wrenches.

ADJUSTABLE WRENCHES

Adjustable wrenches have moveable jaws that are set by adjusting the worm screw on the face of the wrench. Adjustable wrenches come in many different sizes and lengths. The size of the wrench

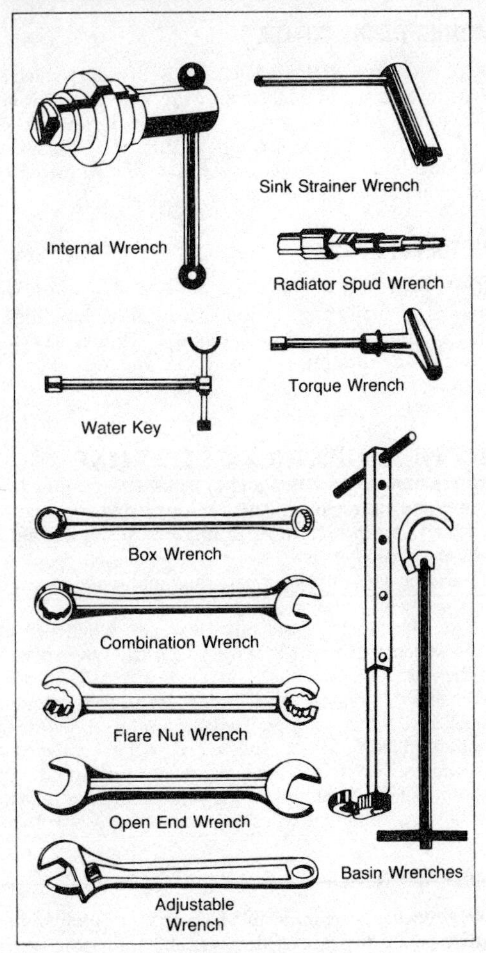

(Figure 1-39) Common types of wrench

is determined by the length of the handle, which determines the width of the jaw opening. The smallest handle length is 4 inches; the largest is 24 inches. The smooth jaws allow you to tighten or loosen many different kinds of plumbing equipment, such as unions and lock nuts. Using an adjustable wrench can save time by avoiding the need to change to a box or open-end wrench when adjusting a variety of bolts.

BASIN WRENCHES

There are two types of basin wrenches. One type comes with a straight shank, and only the head can be turned to loosen or tighten a nut. The other type is adjustable at the head and in the length. By depressing a thumb ball in the handle, you can extend it to various lengths. This wrench is very useful in tight places, such as under kitchen sinks.

BOX WRENCHES

Box wrenches are used to tighten or loosen bolt heads or nuts in tight places. The ends of these wrenches are closed and are placed over the nut or bolt head. Each end of the wrench accommodates a different nut size.

COMBINATION WRENCHES

These wrenches are useful in installing fixtures and in doing repairs. They can be placed over or on the head of a bolt or nut and tightened. Each end of the wrench fits the same nut size.

FLARE-NUT WRENCHES

These wrenches are used on a nut when there is copper tubing attached to it. The open end slips around the copper tubing and onto the flare nut. To tighten, pull the wrench towards you; to loosen, push the wrench away from you.

HEX WRENCHES

Hex wrenches are used to tighten hex heads, such as on extension bits and wood-bit worm screws.

INTERNAL WRENCHES

These wrenches are inserted into pipe nipples and turned to tighten or loosen. They are used when there is not enough room to use a pipe wrench on the outside walls of the nipple.

OPEN-END WRENCHES

These wrenches are useful when installing equipment and for repair work. You need sufficient room to turn the wrench and reposition the open-end part of the wrench on the fitting or nut.

RADIATOR WRENCHES

These wrenches are inserted into the radiator nipple or equipment you wish to remove or tighten. The tapered end allows you to use the wrench on many different sizes of radiator nipples.

SINK-STRAINER WRENCHES

These wrenches hold strainers and keep them from turning while you install or replace them. You can also use them to tighten a strainer into a fitting.

TORQUE WRENCHES

Torque wrenches are used to loosen and tighten flat-headed hex nuts on stainless-steel clamps (no-hub clamps). By turning the wrench, you tighten until a prereleased setting. This prevents you from overtightening and cracking pipes and fittings. To loosen, pull up on the collar and rotate the handle counterclockwise.

WATER KEY, CURB

The water key is placed into the curb box at the property line. Turning the water key one-quarter of a turn will completely shut off the main or turn it on. The crescent-shaped end of the key is placed in the two holes on top of the water box and turned counterclockwise to unscrew or open the curb housing. The other end of the T-bar is used to open another style of curb box, which has an irregular brass nut (vandalproof) that must be unscrewed to open the top and gain access to the housing or hollow stem of the curb box.

WRENCHES, PIPE

Pipe wrenches are used to tighten and loosen all types and sizes of pipe. Some wrenches are used in installation of chrome plumbing materials for "finishing" the plumbing system. Typically, pipe wrenches are durable and can take a lot of use. The jaws need to be replaced when they wear down. Worn or dull pipe jaws can create serious accidents. Figure 1–40 shows many types of pipe wrenches.

CHAIN WRENCHES

Chain wrenches secure the outside wall of the pipe with the chain. Using the handle as leverage, the plumber can install pipe and fittings in tight places. The chain wrench provides a tight grip of 360 degrees on the pipe surface. Chain wrenches are available in sizes from 12 to 36 inches.

END WRENCHES

End wrenches allow an easier grip on pipe and fittings. The jaws of the wrench are adjusted by a nut in the handle. These wrenches can be used in tight places or when working close to a wall. They come in sizes from 6 to 36 inches.

OFFSET HEX WRENCHES

These wrenches can be used in tight places. They have smooth, adjustable jaws. The jaws are smooth so that you don't scratch chrome-finished products, such as lock nuts, flushometers, and chrome unions. The 9½-inch size is the most common.

OFFSET REGULAR WRENCHES

These wrenches are adjustable and are used in tight spaces. They come in sizes from 14 to 24 inches.

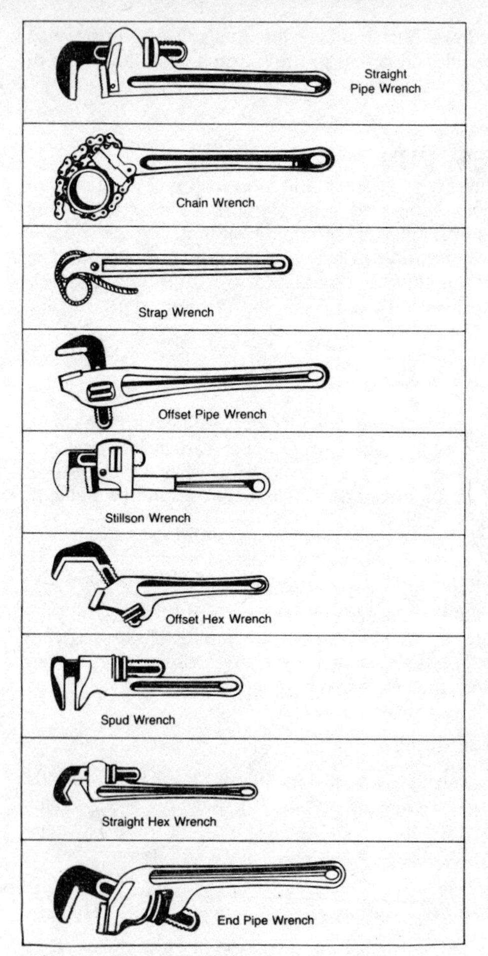

(Figure 1-40) Pipe wrenches

SPUD WRENCHES

Spud wrenches are adjustable and can be used on all types of plumbing equipment. They have narrow jaws to fit into small spaces.

STRAP WRENCHES

Strap wrenches are suitable for any polished brass (chrome) pipe or fittings. They can be used in tight spaces by laying the strap over the pipe and securing it in the handle area. Once the strap has been tightened, you can then use the handle as a lever bar to finish the installation or repair. These wrenches apply equal pressure on the full surface of the area to be tightened or loosened. Strap wrenches are available in handle lengths of 11¾ to 18 inches.

STILLSON WRENCHES

These old-style wrenches are still in use today. They are easy to use, and are most useful on small pipe and fittings. They are adjustable to many pipe sizes by rotating the ring in the handle. They come in sizes from 8 to 24 inches.

STRAIGHT WRENCHES

Straight wrenches are commonly used on pipe and fittings when there is sufficient room to work. They are adjustable and come in many different sizes and lengths. They are durable and versatile.

STRAIGHT HEX WRENCHES

These wrenches are used on flush valves, unions, or any equipment that has a flat surface area. They are adjustable by rotating the ring in the handle. They are used when there is sufficient room to place the wrench on the fitting. Straight hex wrenches are available in sizes from 11¾ to 20 inches.

WRENCHES, REPAIR

These wrenches are used to install and repair various types of equipment in the plumbing system. Plumbers who do repair work will want to add other tools to their toolboxes, but these tools help make repair or installation work easier. See Figure 1–41 for different repair wrenches.

ALL-PURPOSE WRENCHES

These wrenches are used to tighten lock nuts on traps, waste fittings, duo strainers, etc.

BASIN PLUG WRENCHES

These wrenches are used to keep the basin strainer from turning. They are inserted into the middle of the strainer and held tight while turning the lock nut under the basin. This tightens the strainer and keeps it from leaking.

CLOSET SPUD WRENCHES

Closet spud wrenches are used to hold and tighten the lock nuts under the water closet tank assembly.

FAUCET SEAT WRENCHES

These wrenches are inserted into the hot- and cold-water stem areas of a faucet. They remove and replace worn or pitted brass seats in faucets, shower diverters, and bathtub diverters. You tighten the seat by rotating the handle clockwise and loosen it by rotating the handle counterclockwise. Some wrenches have a tapered end, while others have hex-type heads. Care must be taken not to round out the brass seats, which would result in very difficult replacement.

PLUG WRENCHES

Plug wrenches have many uses on bathtub, basins, kitchen sinks, and other equipment. The tool is inserted into the proper hole, and by rotating the opposite handle, you tighten or loosen the plumbing fitting that you are working on.

SOCKET WRENCHES

Socket wrench sets enable you to repair and replace the hot-water or cold-water stems in a faucet. You place the proper-sized socket

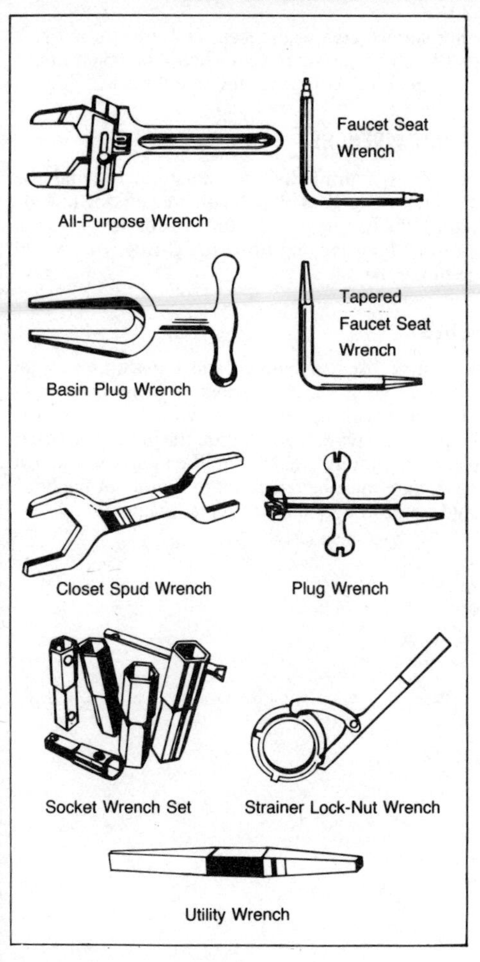

(Figure 1-41) Repair wrenches

over the flat, outer surface area of the stem and turn the wrench handle to loosen or tighten the stem. Care should be taken not to force the stem, or you could easily strip the fine threads.

STRAINER LOCK-NUT WRENCHES

These wrenches are used primarily on duo-strainers (under a kitchen sink) to tighten the large O-ring. If you are working outside the cabinet, you push the handle towards the back of the sink area. If you are lying on the floor looking up at the O-ring, you would turn it clockwise to tighten.

UTILITY WRENCHES

Utility wrenches can be used to remove and replace brass stem washers in the base of faucets and diverters. You insert the tool into the brass stem, and with an adjustable wrench you turn the utility wrench counterclockwise. It will loosen the brass seat. Insert a new seat by placing the utility wrench into the middle of the seat and turn clockwise. Care must be taken not to round out the brass seats, which could result in very difficult replacement.

2

Pipes and Fittings

Steel, cast iron, brass, lead, and copper are the traditional materials used in plumbing. In recent years, fiberglass and various types of plastic have been introduced to the industry so that today pipes and pipe fittings are available in a wide variety of materials.

This chapter discusses the common types of pipe and the fittings used for various purposes with the different types of piping material. Other essential equipment—faucets and valves—are also discussed. The plumber—and apprentice—must be familiar with all these types for repair and alteration work as well as for new installations.

TYPES OF PIPE

Different types of pipes are best suited for particular purposes. A knowledge of the characteristics of the major types of pipe and their uses is most important in choosing the right pipe for a particular job.

BITUMINOUS PIPE (ORANGEBURG)

Bituminous-fiber sewer pipe is sometimes called "Orangeburg" after the leading manufacturer of this type of pipe. Pitch black, bituminous sewer pipe is made of wood fibers and coal tar pitch. (See Figure 2–1.) It has a hard, cardboardlike texture that does not maintain its shape and can be crushed by heavy earth pressures. It is fast being replaced by plastic pipe and may be difficult to find in some hardware or plumbing supply stores. In many states it is illegal to use bituminous pipe in house sewer systems or house drains. It is,

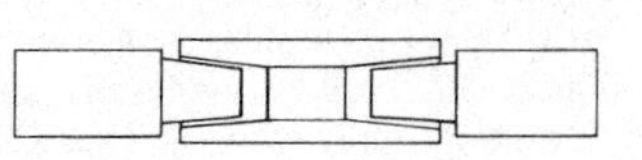

(Figure 2-1) Bituminous (Orangeburg) pipe

TABLE 2–1

SIZES OF ORANGEBURG PIPE	
Inside Diameter (in inches)	Length (in feet)
2	5
3	8
4	8
5	5
6	5

however, still used in septic systems and leaching fields. When considering the use of bituminous pipe or any plumbing material, for that matter, remember to check the local plumbing code for the area you are working in. All plumbing work is governed by code, and you must adhere to it strictly. Table 2–1 lists the sizes of bituminous, or Orangeburg, that are available. This type of pipe can be cut with a hand saw.

To install bituminous fiber pipe, place a coupling or other type of fitting on the tapered end of the pipe. Then force the joints together by leverage or by striking a block of wood against the end.

BRASS PIPE

Brass is a yellowish alloy composed of copper and zinc in different proportions; red brass, with a higher proportion of copper, is most often found in piping systems. Brass pipe can be used on steam systems and on potable water-supply systems.

Brass is very durable and versatile and does not rust or corrode easily. It is much more expensive than other types of pipe, and so it is not as widely used now as it used to be. However, most plumbing supply stores still carry brass pipe and fittings.

Brass pipe is made in sizes ranging from ⅛-inch to 12-inch (same sizes as for steel pipe) and usually comes in 21-foot lengths. Long pipe sizes can withstand pressures of 1,000 psi. (See Figure 2–2.)

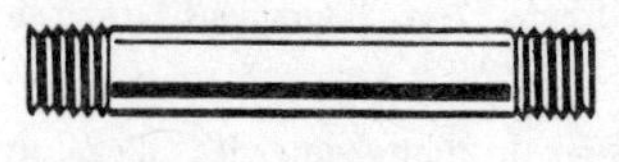

(Figure 2-2) Brass pipe

Brass pipe is soft, so threads can be made quite easily, but be careful not to overtighten because the threads can be stripped with even small effort. Tightening may be done with a pipe wrench, strap wrench, or spud wrench, depending on the brass finish.

Chrome-Plated Brass Piping. Chrome-plated brass is used on fixtures to supply hot and cold water to shower stalls, kitchen equipment, and other fixtures. It is a plumbing installation that is pleasing to the eye.

Regular Brass Piping. Regular brass is used on steam lines and runs of hot- and cold-water lines where it is not necessary to use chrome plate for appearance.

CAST-IRON SOIL PIPE

Cast iron is an alloy of iron, carbon, and silicon. There are many types of cast iron, gray being the most common. Cast-iron soil pipe is used for drainage, sewer, and venting systems. It is very strong, but heavy and bulky, usually requiring two persons to lift and install it. It was widely used in the past but is now being replaced by lighter-weight pipes, such as plastic.

There are several types of cast-iron soil pipe: pipe with beaded ends, pipe with bell and spigot ends, and no-hub pipe.

BEADED-END CAST-IRON SOIL PIPE

Cast-iron soil pipe with beaded ends is made in various sizes: 2-inch, 3-inch, 4-inch, 5-inch, 6-inch, 8-inch, 10-inch, 12-inch, and 15-inch inside diameter. There are two weights: service weight and extra-heavy weight for long wear and usage. These pipes are joined with lead and oakum joints to make them airtight and watertight. Tables 2–2 and 2–3 provide data on the length, weight, and other dimensions of beaded pipe.

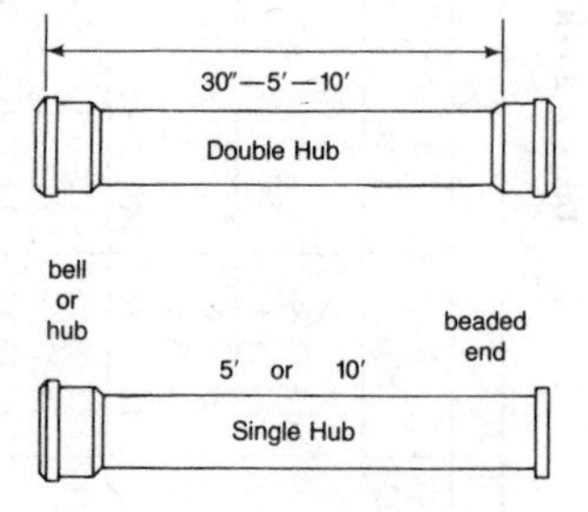

(Figure 2-3) Beaded end cast iron pipe

Beaded cast-iron pipe comes in double-hub and single-hub forms (See Figure 2–3.)

TABLE 2–2 LENGTH DATA (IN INCHES) FOR BEADED CAST-IRON PIPE

EXTRA-HEAVY PIPE 'XH'								
Size (Nom. I.D.)	K Max.	H Max.	J	F	Y	E	M	N
2	4 1/8	3 5/8	2 3/8	3/4	2 1/2	2 3/4	2 3/4	11/16
3	5 3/8	4 15/16	3 1/2	13/16	2 3/4	3 1/4	3 7/8	3/4
4	6 3/8	5 15/16	4 1/2	7/8	3	3 1/2	4 7/8	13/16
5	7 3/8	6 15/16	5 1/2	7/8	3	3 1/2	5 7/8	13/16
6	8 3/8	7 15/16	6 1/2	7/8	3	3 1/2	6 7/8	13/16
8	11 1/16	10 7/16	8 5/8	1 3/16	3 1/2	4 1/8	9	1 1/8
10	13 5/16	12 11/16	10 3/4	1 3/16	3 1/2	4 1/8	11 1/8	1 1/8
12	15 7/16	14 13/16	12 3/4	1 7/16	4 1/4	5	13 1/8	1 3/8
15	18 13/16	18 3/16	15 7/8	1 7/16	4 1/4	5	16 1/4	1 3/8

SERVICE PIPE 'SV'								
Size (Nom. I.D.)	K Max.	H Max.	J	F	Y	E	M	N
2	3 15/16	3 3/8	2 1/4	3/4	2 1/2	2 3/4	2 5/8	11/16
3	5	4 1/2	3 1/4	13/16	2 3/4	3 1/4	3 5/8	3/4
4	6	5 1/2	4 1/4	7/8	3	3 1/2	4 5/8	13/16
5	7	6 1/2	5 1/4	7/8	3	3 1/2	5 5/8	13/16
6	8	7 1/2	6 1/4	7/8	3	3 1/2	6 5/8	13/16
8	10 1/2	9 7/8	8 3/8	1 3/16	3 1/2	4 1/8	8 3/4	1 1/8
10	12 13/16	12 3/16	10 1/2	1 3/16	3 1/2	4 1/8	10 7/8	1 1/8
12	14 15/16	14 5/16	12 1/2	1 7/16	4 1/4	5	12 7/8	1 3/8
15	18 5/16	17 5/8	15 5/8	1 7/16	4 1/4	5	16	1 3/8

COURTESY OF CAST IRON SOIL PIPE INSTITUTE

TABLE 2–3

WEIGHT OF SOIL PIPE

Service			
Size	**Lbs. per Ft.**	**Size**	**Lbs. per Ft.**
2″	4	6″	15
3″	6	7″	20
4″	9	8″	25
5″	12		

Extra Heavy			
Size	**Lbs. per Ft.**	**Size**	**Lbs. per Ft.**
2″	5	8″	30
3″	9	10″	43
4″	12	12″	54
5″	15	15″	75
6″	19		

COURTESY OF CAST IRON SOIL PIPE INSTITUTE

Double-Hub Beaded Cast-Iron Pipe. This type of pipe is used to reduce pipe waste. Usually 5-foot 30-inch lengths are cut to avoid waste associated with cutting 5-foot or 10-foot lengths of single-hub pipe. After cutting this type of pipe, the plumber still has a hub or bell to use in some other area. It is illegal to use double-hub cast-iron lengths without cutting them.

Single-Hub Beaded Cast-Iron Pipe. This type of pipe is used for long runs, either horizontally or vertically.

BELL AND SPIGOT CAST-IRON SOIL PIPE

Cast-iron soil pipe with bell and spigot ends is used, like beaded-end pipe, to convey waste and air for venting systems. It comes in the same sizes—2-inch through 15-inch inside diameter—as beaded pipe, the same two weights—service and extra-heavy—and the same two forms—double-hub and single-hub.

The ends of the pipe are smooth and allow easy entry for a rubber gasket connection. Pipes can also be joined with lead and oakum joints. (See Figure 2–4.)

Double Hub, 5′ Lengths									
Diameter	2″	3″	4″	5″	6″	8″	10″	12″	15″
SV Weight	21	31	42	54	68	105	150	200	270
XH Weight	26	47	63	78	100	157	225	285	395

Double Hub, 30″ Lengths		
Diameter	SV Weight	XH Weight
2″	11	14
3″	17	26
4″	23	33

Single Hub, 5′ Lengths									
Diameter	2″	3″	4″	5″	6″	8″	10″	12″	15″
SV Weight	20	30	40	52	65	100	145	190	255
XH Weight	25	45	60	75	95	150	215	270	375

Single Hub, 10′ Lengths		
Diameter	SV Weight	XH Weight
2″	38	43
3″	56	83
4″	75	108
5″	98	133
6″	124	160
8″	185	265
10″	270	400
12″	355	480
15″	475	705

(Figure 2-4) Bell and spigot cast iron pipe

Double-Hub Bell and Spigot Cast-Iron Soil Pipe. This type of pipe is used to reduce the waste associated with cutting a long length of pipe.

Single-Hub Bell and Spigot Cast-Iron Soil Pipe. This type of pipe is used to cover long distances, either vertically or horizontally.

NO-HUB CAST-IRON SOIL PIPE

No-hub soil pipe is used in vent and drainage systems, often on high-rise apartment buildings and commercial buildings. It is made

No-Hub Pipe, 10′ Lengths							
Diameter	1-1/2	2	3	4	5	6	8
Weight per 10′	27	38	54	74	95	118	180

(Figure 2-5) No-hub cast iron pipe

in sizes from 1½-inch to 10-inch inside diameter, in 10-foot lengths, and only in service weight. All no-hub pipe and fittings are labelled with the manufacturer's name, the Cast Iron Soil Pipe Institute's trademark, and the size. The label is about 1½ inches from the coupling joint. (See Figure 2–5.)

No-hub pipe comes with tapered ends only. It can be cut to various lengths with a cast-iron chain cutter, a saw with a special blade, a hacksaw, or a hammer and chisel. Lengths of pipe are joined using a mechanical joint (neoprene sleeve with a stainless steel clamp assembly). Lead and oakum joints should not be used.

COPPER TUBING

Copper tubing has been the staple of plumbing for many years. It is used in drainage and venting systems as well as in potable water-supply systems. It is extremely durable, and a carefully done installation will last for many years. The tubing may be flared, soldered, and tightened with compression fittings to prevent leaks. It is not easily welded, but it can easily be brazed. (See Figure 2–6.)

(Figure 2-6) Copper tubing

TYPES OF COPPER TUBING

Copper tubing comes in two main types—hard (drawn) and soft (annealed). Table 2–4 lists the diameters and lengths available in the different types.

Drawn (Hard) Copper Tubing. Drawn copper is straight and rigid. It is easily installed, and its rigidity gives a professional

TABLE 2–4

AVAILABLE LENGTHS OF COPPER PLUMBING TUBE

TUBE	DRAWN (hard copper)		ANNEALED (soft copper)	
Type K	Straight Lengths:		Straight Lengths:	
	Up to 8-inch diameter	20 ft	Up to 8-inch diameter	20 ft
	10-inch diameter	18 ft	10-inch diameter	18 ft
	12-inch diameter	12 ft	12-inch diameter	12 ft
			Coils:[1]	
			Up to 1-inch diameter	60 ft
				100 ft
			1¼ and 1½-inch diameter	60 ft
			2-inch diameter	40 ft
				45 ft
Type L	Straight Lengths:		Straight lengths:	
	Up to 10-inch diameter	20 ft	Up to 10-inch diameter	20 ft
	12-inch diameter	18 ft	12-inch diameter	18 ft
			Coils:	
			Up to 1-inch diameter	60 ft
				100 ft
			1¼ and 1½-inch diameter	60 ft
				100 ft
			2-inch diameter	40 ft
				45 ft
Type M	Straight Lengths:		Not available	—
	All diameters	20 ft		
DWV	Straight Lengths:		Not available	—
	All diameters	20 ft		

[1] for water service

COURTESY OF COPPER DEVELOPMENT ASSOCIATION, INC.

appearance over long distances. It is available in all types of tubes (see below) for different uses, in lengths of 12, 18, or 20 feet, and in diameters from ¼ inch to 12 inches. The lengths are coded and stamped with a specific color identifying the type of tubing. Hard tubing is usually soldered onto copper fittings.

Annealed (Soft) Copper Tubing. Annealed copper is flexible. It comes in coils of 30, 40, 60, or 100 feet and always retains its curve. It is used when many turns and bends are necessary.

Annealed copper is only available in certain types of tubes (in terms of wall thicknesses and uses). It has many uses but is used primarily to run potable water underground. There is no color code on the coil, but a code is stamped into the tubing for identification. Coils of copper tubing are usually flared or soldered during installation. Two pipe wrenches should be used when tightening flared ends and fittings.

TYPES OF TUBES

There are four types of copper tubes—K, L, M, and D.W.V.; they differ in wall thickness, weight per foot, bursting pressure, and safe working stress—and therefore in their proper uses. Table 2–5 lists the characteristics of each type of tube.

K Type. Type K tube, available in both drawn and annealed tubing, is the strongest. It is used for underground water service and general plumbing and heating installations.

L Type. Type L tube, available in both drawn and annealed form, is used for general plumbing and heating installations. It is usually used to convey hot and cold potable water to fixtures.

M Type. Type M tube, available only in drawn tubing, is used for general plumbing, heating, and drainage installations. It is most commonly used for hot-water baseboard, radiator, and heating systems.

D.W.V. Type. Drain, waste, and vent (D.W.V.) tube, also available only in rigid form, can be used for drain, waste, vent, and other nonpressure applications above ground—if allowed by local code. (Some states do not permit the use of D.M.V. copper tubing.) Its most common use is for rain downspouts.

PLASTIC PIPING

Plastic piping is used for waste and vent piping as well as for carrying hot and cold water. It usually comes in 20-foot lengths, which are cemented together with appropriate material, depending on the type of plastic tubing.

There are several types of plastic tubing, including ABS, CPVC, PVC, and PE.

TABLE 2–5

COPPER TUBING DATA

Nominal Tube Size in Inches	Outside Diameter in Inches	TOLERANCE Tubing O.D. In Inches		TYPE "K" — Hard Drawn—20 Ft. Straight Lengths Soft Annealed—20 Ft. Straight Lengths or 40 Ft. and 60 Ft. Length Coils (to and including 1½") — Use: For Underground Service and General Plumbing and Heating Installations under Severe Conditions						TYPE "L" — Hard Drawn—20 Ft. Straight Lengths Soft Annealed—20 Ft. Straight Lengths or 40 Ft. and 60 Ft. Length Coils (to and including 1½") — Use: For General Plumbing and Heating Installations					
						HARD DRAWN		SOFT ANNEALED				HARD DRAWN		SOFT ANNEALED	
		Max.	Min.	Wall Thickness In Inches	Weight per Foot in Lbs.	Bursting Pressure in Lbs.*	Safe Working Stress in Lbs.⊗●	Bursting Pressure in Lbs.*	Safe Working Stress in Lbs.☆●	Wall Thickness In Inches	Weight per Foot in Lbs.	Bursting Pressure in Lbs.*	Safe Working Stress in Lbs.⊗●	Bursting Pressure in Lbs.*	Safe Working Stress in Lbs.★●
¼"	.375	.376	.374	.035	.145	6720	1060	5600	930	.030	.126	5760	900	4800	800
⅜"	.500	.501	.499	.049	.269	7100	1170	5900	980	.035	.196	5000	800	4200	700
½"	.625	.626	.624	.049	.344	5600	920	4700	780	.040	.285	4600	740	3800	630
⅝"	.750	.751	.749	.049	.418	4700	760	3900	650	.042	.362	4000	650	3400	560
¾	.875	.876	.874	.065	.641	5300	880	4500	750	.045	.455	3700	590	3100	510
1"	1.125	1.1265	1.1235	.065	.839	4200	680	3500	580	.050	.655	3200	510	2700	450
1¼"	1.375	1.3765	1.3735	.065	1.04	3400	550	2800	465	.055	.884	2900	460	2400	500
1½"	1.625	1.627	1.623	.072	1.36	3200	520	2700	450	.060	1.14	2700	430	2200	360
2"	2.125	2.127	2.123	.083	2.06	2800	450	2300	380	.070	1.75	2400	370	2000	330

Nominal Tube Size in Inches	Outside Diameter in Inches	TOLERANCE Tubing O.D. In Inches		TYPE "K"						TYPE "L"					
						HARD DRAWN		SOFT ANNEALED				HARD DRAWN		SOFT ANNEALED	
		Max.	Min.	Wall Thickness In Inches	Weight per Foot in Lbs.	Bursting Pressure in Lbs.*	Safe Working Stress in Lbs.⊗●	Bursting Pressure in Lbs.*	Safe Working Stress in Lbs.☆●	Wall Thickness In Inches	Weight per Foot in Lbs.	Bursting Pressure in Lbs.*	Safe Working Stress in Lbs.⊗●	Bursting Pressure in Lbs.*	Safe Working Stress in Lbs.★●
2½"	2.625	2.627	2.623	.095	2.92	2600	420	2200	360	.080	2.48	2220	350	1800	300
3"	3.125	3.127	3.123	.109	4.00	2500	410	2100	350	.090	3.33	2100	330	1700	280
3½"	3.625	3.627	3.623	.120	5.12	2400	380	2000	330	.100	4.29	2000	320	1700	280
4"	4.125	4.127	4.123	.134	6.51	2300	370	1900	320	.110	5.38	1900	300	1600	260
5"	5.125	5.127	5.123	.160	9.67	2200	360	1900	320	.125	7.61	1800	280	1500	250
6"	6.125	6.127	6.123	.192	13.87	2300	370	1900	320	.140	10.20	1600	260	1400	230
8"	8.125	8.127	8.121	.271	25.90	2400	390	2000	330	.200	19.29	1800	280	1500	250
10"	10.125	10.127	10.119	.338	40.3	2400	390	2000	330	.250	30.1	1800	290	1500	250
12"	12.125	12.127	12.119	.405	57.8	2400	400	2000	330	.280	40.4	1700	270	1400	230

Note: Information and data contained in these charts as taken from A.S.T.M. Specifications No. B-88-72, Federal Specification WW-T-799, and various Copper Tube Mill chart standards.

⊗ From Cabra based on 150°F. with an allowable stress of 6000 P.S.I.

* Bursting pressures are calculated from the following Formula for thin, hollow cylinders under tension:

$$P = \frac{2tS}{D}$$

Where P = Bursting pressure, Lb. per Sq. In.
t = Wall thickness, inches
D = Outside tube diameter, inches
S = Tensile strength (36,000 Lb. per Sq. In. for hard tubes and 30,000 for soft tubes)

☆ With safety factor of 6, maximum safe working pressure allowable by common usage up to 150°F, can be taken at 1/6 the above bursting pressure

● Rated internal pressure for copper water tube based on the strength of the tube alone and applicable to systems using suitable mechanical joints. (Pounds per Sq. Inch).

Nominal Tube Size in Inches	Outside Diameter in Inches	TOLERANCE Tubing O.D. In Inches		TYPE "M" Hard Drawn Only—20 Ft. Straight Lengths. Use: For General Plumbing, Heating and Drainage Installations				TYPE "DWV" Hard Drawn Only—20 Ft. Straight Lengths. Use: For Drain, Waste, Vent and other non-pressure applications above ground.		
						HARD DRAWN				HARD DRAWN
		Max.	Min.	Wall Thickness In Inches	Weight per Foot in Lbs.	Bursting Pressure in Lbs.*	Safe Working Stress in Lbs.⊗●	Wall Thickness in Inches	Weight per Foot in Lbs.	Bursting Pressure in Lbs.
3/8"	.500	.501	.499	.025	.144	3600	560	....	....	....
1/2"	.625	.626	.624	.028	.203	3230	510	....	....	....
3/4"	.875	.876	.874	.032	.328	2640	420	....	....	....
1"	1.125	1.1265	1.1235	.035	.464	2240	340	....	....	....
1 1/4"	1.375	1.3766	1.3736	.042	.681	2220	340	.040	.650	1950
1 1/2"	1.625	1.627	1.623	.049	.940	2170	340	.042	.809	1740
2"	2.125	2.127	2.123	.058	1.46	1965	300	.042	1.07	1310
2 1/2"	2.625	2.627	2.623	.065	2.03	1780	280	....	....	....
3"	3.125	3.127	3.123	.072	2.68	1660	260	.045	1.69	950
3 1/2"	3.625	3.627	3.623	.083	3.58	1640	260	....	....	....
4"	4.125	4.127	4.123	.095	4.66	1650	260	.058	2.87	900
5"	5.125	5.127	5.123	.109	6.66	1530	240	.072	4.43	910
6"	6.125	6.127	6.123	.122	8.91	1430	230	.083	6.10	890
8"	8.125	8.127	8.123	.170	16.46	1510	240	....	....	....
10"	10.125	10.127	10.119	.212	25.6	1510	240	....	....	....
12"	12.125	12.127	12.119	.254	36.7	1510	240	....	....	....

COURTESY OF COPPER DEVELOPMENT ASSOCIATION, INC.

ABS PIPE

Acrylonitrile–butadiene–styrene (ABS) pipe is used in irrigation lines, drain lines, and for waste and vent piping. It can be used at temperatures up to 180° F. The pipe is black and rigid. It comes in 20-foot lengths in a variety of diameters from 1¼ inches to 6 inches. Pipe known as Schedule #40 is usually used for drainage and venting. ABS pipe can be cut with tubing cutters or a hand saw. Pipe and fittings are welded together using special ABS cement. (See Figure 2–7a.)

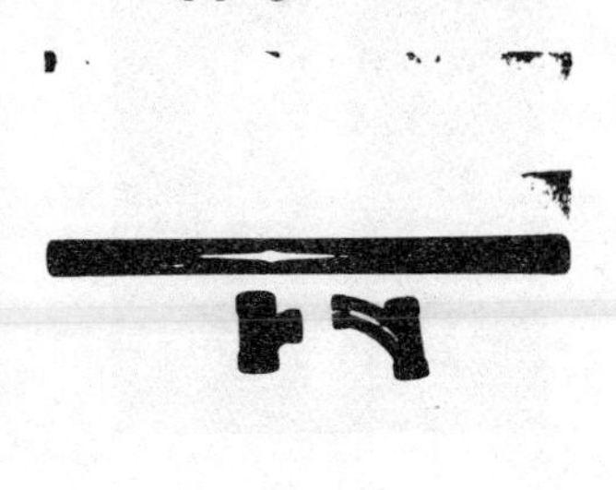

(Figure 2-7a) ABS pipe

CPVC PIPE

Chlorinated polyvinylchloride (CPVC) pipe is used on hot- and cold-water distribution lines. It is designated for water service up to and including 180° F. White or gray, the rigid pipe is most commonly used in ½-inch and ¾-inch sizes. One of its common uses is in underground and overhead sprinkler systems—in gardens and greenhouses, for example. It can be cut using a hand saw or a tubing cutter. CPVC cannot be used in all areas, so be sure to check the local code. (See Figure 2–7b.)

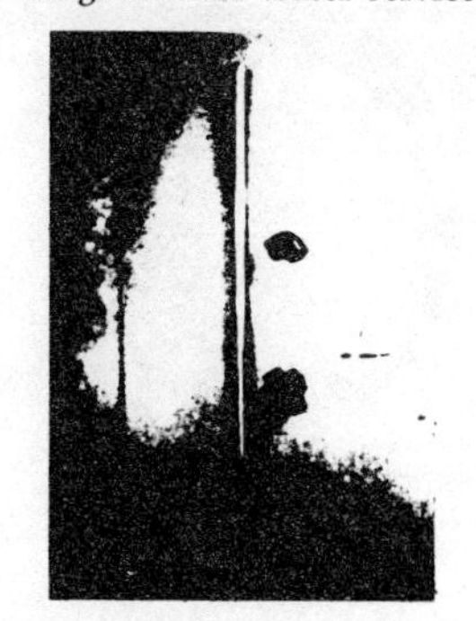

(Figure 2-7b) CPVC pipe

PVC PIPE

Polyvinylchloride (PVC) pipe is used in installing waste and vent lines. This rigid white pipe is available in many sizes, ranging from 1¼ inch to 6 inches and

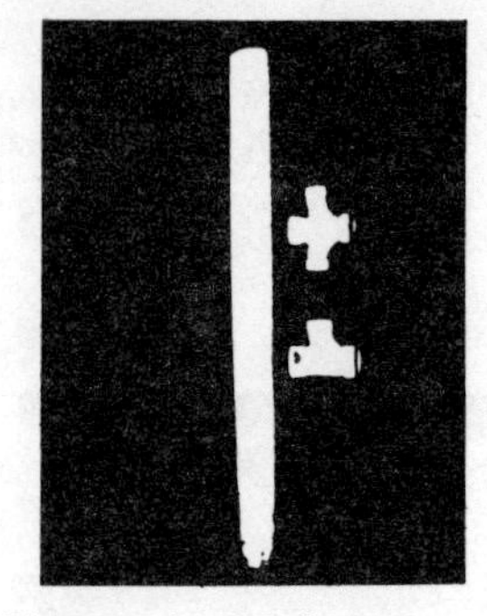

(Figure 2-7c) PVC pipe

comes in 20-foot lengths. Schedule #40 is usually used on plumbing and drainage systems. The pipe is cut with a hand saw or tubing cutter; pipe and fittings are cemented together with PVC cement. (See Figure 2–7c.)

PE Pipe

Polyethylene (PE) tubing is primarily used to carry potable water from a well or street water main. It is also used on irrigation lines. A flexible type of tubing, it is available in ¾-inch, 1-inch, 1½-inch, and 2-inch sizes. It comes in coils up to 200 feet.

(Figure 2-7d) PE pipe

PE coils are usually unrolled in a ditch and joined with PE fittings and stainless steel clamps. It can also be flared using the proper flaring tool. When unrolling PE tubing, be careful not to kink or bend it. A kink can result in a decrease in water volume and eventually in pressure. If a kink should develop, cut it out and join the ends of the tubing with a coupling. (See Figure 2–7d.)

STEEL PIPE

Steel pipe, especially galvanized steel pipe, is by far the most commonly used type of pipe. It comes in 21-foot lengths with interior diameters from ⅛-inch to 12 inches. Large sizes (above 1¼-inch diameter) are difficult to handle, and two workers are usually needed to move large amounts of this pipe. All steel pipe comes with factory cut threads (ASPT) on both ends.

TABLE 2–6

DIFFERENCES BETWEEN STANDARD AND EXTRA-HEAVY-WEIGHT STEEL PIPE

		SCHEDULE 40 (standard wall)		SCHEDULE 80 (extra strong wall)	
Nominal Pipe Size (inches)	Outside Diameter (Inches)	Wall Thickness (inches)	Inside Diameter (inches)	Wall Thickness (inches)	Inside Diameter (inches)
⅛	.405	.068	.269	.095	.215
¼	.540	.088	.364	.119	.302
⅜	.675	.091	.493	.126	.423
½	.840	.109	.622	.147	.546
¾	1.050	.113	.824	.154	.742
1	1.315	.133	1.049	.179	.957
1¼	1.660	.140	1.380	.191	1.278
1½	1.900	.145	1.610	.200	1.500
2	2.375	.154	2.067	.218	1.939
2½	2.875	.203	2.469	.276	2.323
3	3.500	.216	3.068	.300	2.900
3½	4.000	.226	3.548	.318	3.364
4	4.500	.237	4.026	.337	3.826
5	5.563	.258	5.047	.375	4.813
6	6.625	.280	6.065	.432	5.761
8	8.625	.322	7.981	.500	7.625
10	10.750	.365	10.020	.500 (xs)	9.750
12	12.750	.375 (std)	12.000	.500 (xs)	11.750
		.406 (sch 40)	11.938		

COURTESY OF THE STEEL PIPE INSTITUTE

Long lengths of steel pipe are made in three weights: standard, extra-heavy, and double extra-heavy. Most plumbers use standard weight steel pipe, but in some cases stronger pipes must be used. Table 2–6 lists the differences in wall thickness and inside diameter of widely used standard-weight pipe and extra-strong-weight pipe. Notice that the inside diameter decreases as the wall thickness increases. However, since the outside

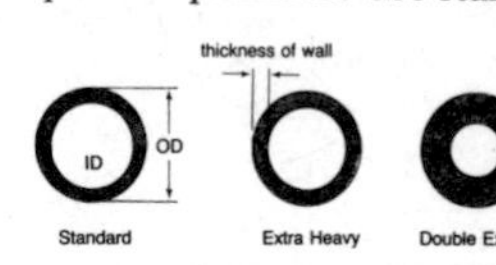

(Figure 2-8) Types of steel pipe

diameter does not change, the same threading dies fit all weights of steel pipe. (See Figure 2–8.)

TYPES OF STEEL PIPE

There are two main types of steel pipe: black pipe and galvanized pipe. A third type, solderable galvanized pipe, has recently been introduced.

Black Pipe. Black pipe does not have a zinc coating. It is not used on water lines but instead on air lines, water-heating systems, and natural gas lines. Its main advantage is its low cost compared to other types of pipe.

Galvanized Steel Pipe. Galvanized steel pipe is coated with zinc on the inside and outside to protect it from rusting quickly. Galvanized steel pipe is used in drainage systems, potable water lines, air lines, vent lines, and natural gas lines.

TYPES OF WELDS ON STEEL PIPE

Steel pipe may be seamless or have one of two types of seams.

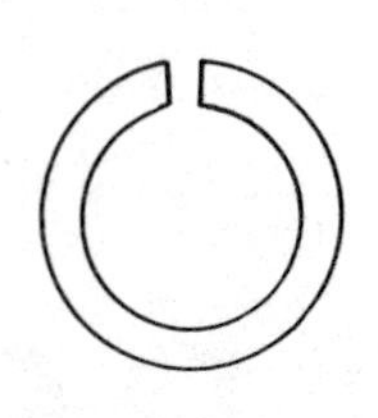

(Figure 2-9a) Butt weld

Butt-Weld Pipe. Butt-weld pipe has the edges cut straight. It is the weakest of the two types of welds because the seam of the weld is short. (See Figure 2–9a.)

Lap-Weld Pipe. Lap-weld pipe has edges cut on a slant. This allows greater welding surface—and thus greater strength. (See Figure 2–9b.)

Seamless Pipe. Seamless pipe has no welded joints at all. It is made by drawing hot steel through a forming machine. It is the strongest type of pipe. (See Figure 2–9c.)

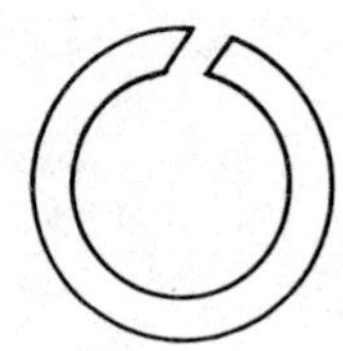

(Figure 2-9b) Lap weld

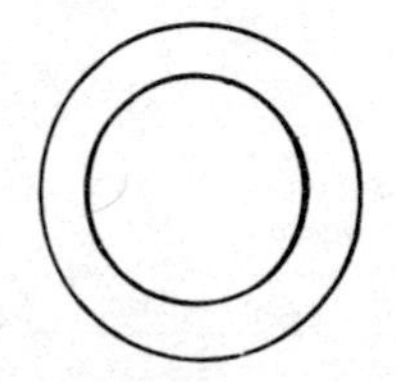

(Figure 2-9c) Seamless

DETERMINING A PIPE'S SAFE WORKING PRESSURES

Butt-weld and lap-weld steel pipe of the same size differ in the amount of pressure they can withstand. Table 2–7 lists the test pressures for three different weights of many common sizes of pipe.

The formula for determining safe working pressures is

$$BP = \frac{2T \times TS}{OD}$$

in which BP is the bursting pressure in pounds per square inch (psi),

T is the thickness of the wall in inches,

OD is the outside diameter of the pipe in inches, and

TS is the tensile strength: 40,000 psi for butt-welded pipe, 50,000 psi for lap-welded pipe.

PIPE FITTINGS

Fittings are the parts of a piping system that join two lengths of pipe. They are essential to any piping system and are used when rigid pipe must be turned, when a pipe branches into two or more pipes, and when any change in size or material is needed.

TABLE 2–7

STEEL PIPE TEST PRESSURES

STANDARD			EXTRA STRONG			DOUBLE EXTRA STRONG		
	Test pressure in pounds			Test pressure in pounds			Test pressure in pounds	
Size	Butt	Lap	Size	Butt	Lap	Size	Butt	Lap
⅛	700		⅛	700				
¼	700		¼	700				
⅜	700		⅜	700				
½	700		½	700		½	700	
¾	700		¾	700		¾	700	
1	700		1	700		1	700	
1¼	700	1000	1¼	1500		1¼	2200	
1½	700	1000	1½	1500	2500	1½	2200	3000
2	700	1000	2	1500	2500	2	2200	3000
2½	800	1000	2½	1500	2000	2½	2200	3000
3	800	1000	3	1500	2000	3		3000
3½		1000	3½		2000	3½		2500
4		1000	4		2000	4		2500
4½		1000	4½		1800	4½		2000
5		1000	5		1800	5		2000
6		1000	6		1800	6		2000
7		1000	7		1500	7		2000
8		800	8		1500	8		2000
8		1000				8		
9		900	9		1500	9		
10		600	10		1200	10		
10		800				10		
10		900				10		
11		800	11		1100	11		
12		600	12		1100	12		
12		800				12		
13		700	13		1000	13		
14		700	14		1000	14		
15		600	15		1000	15		

COURTESY OF THE STEEL PIPE INSTITUTE

There are many types of fittings. They differ in material, such as cast iron, copper, or plastic. They also differ in specific purpose. Some, for example, may be designed for use when an increase or decrease in pipe size is desired, others for use when pipes of two different materials are to be joined. Fittings may also be differentiated by the type of service they are intended for, such as potable water, drainage, and venting.

This section categorizes fittings in two ways: by the type of fitting (adapter, bushing, nipple, etc.) and by the material and type of pipe the fitting is used on (copper, steel, plastic, etc.).

ADAPTERS

Adapters are used to join pipes of different materials (for example, galvanized and copper) or of different sizes. They are used for pipe installation and in alteration and repair work. There are many types of adapters, some copper, some plastic. (See Figure 2–10.)

COPPER ADAPTERS

Copper adapters are used to continue a piping system with copper tubing and piping. They are connected to cast-iron, galvanized, or brass piping to achieve the change in material.

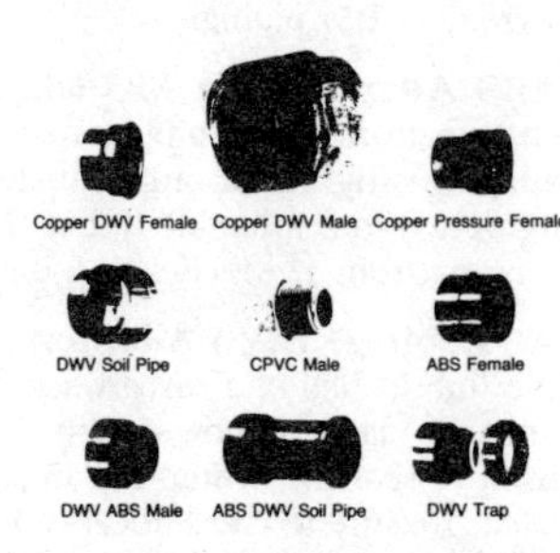

(Figure 2-10) Adapters join pipes of different materials or size

D.W.V. Adapters. Drain, waste, and vent (D.W.V.) adapters are used on drainage piping to continue the waste system with copper tubing and fittings. A male adapter is screwed into an iron-pipe-sized (I.P.S.) fitting; a female adapter is simply tightened onto male threads. D.W.V. adapters come in sizes from 1¼ inches to 4 inches.

Pressure Adapters. A pressure adapter is used to change or add on to a galvanized or brass potable water system. A female fitting is tightened onto male threads and then attached to copper tubing; a male adapter is then screwed into the female fitting. A pressure adapter has a deeper soldering surface inside than a D.W.V. fitting.

This surface allows the adapter to be used on high water-pressure systems, including hot-water heating systems. Fitting sizes range from ⅜ inch to 6 inches.

Soil Adapters. A soil adapter is placed in the hub or bell of a cast-iron soil pipe to allow the plumber to continue the waste piping system with copper or plastic tubing and fittings. Lead and oakum are usually used to make this type of fitting airtight and watertight. However, some state plumbing codes allow the plumber to insert a soil adapter into a rubber gasket. Remember to check the code for the local area. Soil adapters are available in many sizes, generally ranging from 2 by 1¼ inches up to 4 inches.

PLASTIC ADAPTERS

Plastic adapters are used to continue a piping system with chlorinated polyvinyl chloride (CPVC) piping or acrylonitrile–butadiene–styrene (ABS) piping.

ABS Adapters. An ABS adapter is used in a waste system to change from other piping materials to ABS piping. The female adapter is tightened onto a male thread; the male adapter is then tightened into a female fitting. These fittings are available in sizes ranging from 1¼ inches to 6 inches.

ABS Soil (D.W.V.) Adapters. An ABS soil adapter is placed in the hub or bell of a cast-iron soil pipe to change from cast iron to an ABS plastic piping system. Lead and oakum joints are usually used to seal the fitting, but in some states rubber gaskets may be used. Be sure to check local codes. ABS soil adapters are normally sized from 1½ inches to 4 inches.

ABS Trap Fittings (D.W.V.). An ABS trap fitting is used to seal brass traps under a kitchen sink, basin, bathtub, or similar fixture. The brass trap is inserted into the compression area of the fitting. A lock nut and washer are then tightened to make the connection airtight and watertight. ABS trap fittings come in sizes from 1¼ inches to 2 inches.

CPVC Adapters. CPVC adapters change a piping system to CPVC pipe. A male adapter is tightened into a female fitting. Fittings normally come in ½-inch and ¾-inch sizes.

BITUMINOUS FIBER SEWER (ORANGEBURG) FITTINGS

Bituminous fiber sewer fittings are used on bituminous pipe to change the direction of flow of waste water. They are available in 2-inch, 3-inch, 4-inch, 5-inch, and 6-inch sizes. One end of the fitting is tapered; the other has a bell or hub. The fittings are pressured or driven together. Several types are available. (See Figure 2–11.)

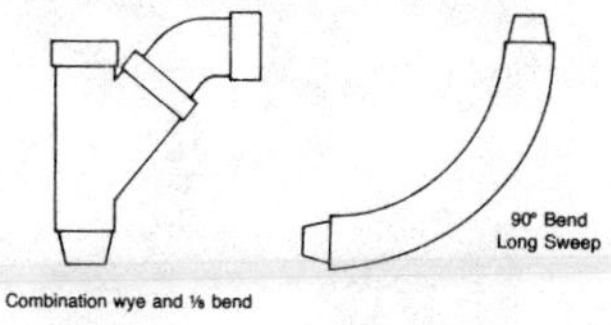

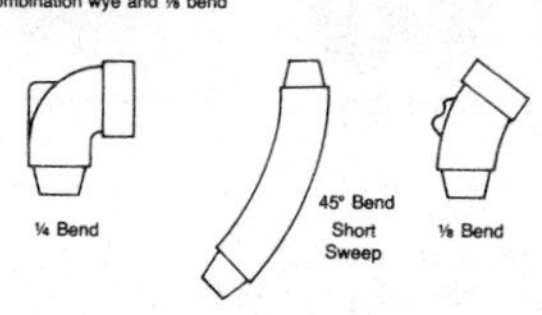

(Figure 2-11) Fittings for bituminous fiber (sewer) pipe

Combination Wye and ⅛ Bends. A combination wye and ⅛ bend is used to combine two waste lines or as a cleanout access area in a vertical position.

Long Sweeps. A long sweep has two tapered ends and is used to change the direction of Orangeburg pipe. It changes the direction of waste water flow gradually.

⅛ Bends. A ⅛ bend is used to change the direction of Orangeburg pipe. It can be used with other fittings.

¼ Bends. A ¼ bend is used to change the direction of Orangeburg pipe by 90 degrees.

Short Sweeps. A short sweep has two tapered ends and is used to change the direction of Orangeburg pipe.

BLACK STEEL FITTINGS, MALLEABLE IRON

Black steel fittings of malleable iron are made by annealing cast iron and are very strong. The standard fitting can withstand working pressures of up to 150 psi. Black steel fittings are used on natural gas lines, air lines, and propane gas installations. Since they are not coated with zinc, they cannot be used on potable water-supply systems. The fitting is simply turned onto a male thread and tightened with a pipe wrench.

These fittings are quite common and can be found in most plumbing supply and hardware stores. Several types of elbows and other fittings are available, in sizes from ⅛ inch to 6 inches. (See Figure 2–12.)

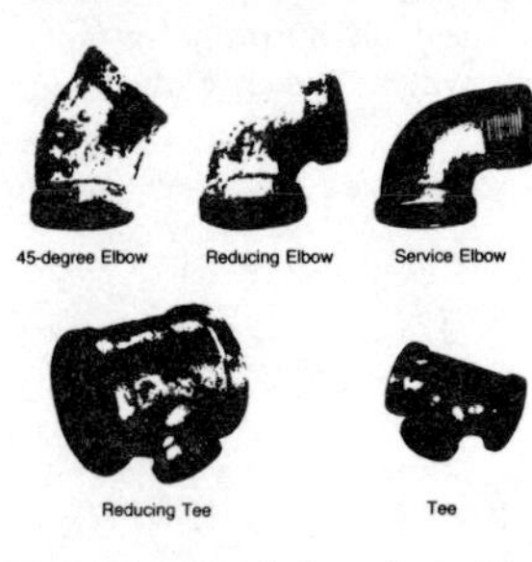

(Figure 2-12) Black steel pipe fittings

45-Degree Elbows. A 45-degree elbow is used to change the direction of a piping system by 45 degrees, usually to avoid obstructions vertically or horizontally.

Reducing Elbows. A reducing elbow is used to reduce or increase pipe sizes.

Service Elbows. A service elbow, also known as a *street elbow,* is used to make swing joints or in places where two elbows and a close nipple cannot be used.

Tees. A tee is a type of fitting that joins a branch pipe to a main pipe at a right angle. Tee sizes are given by listing the straight-through dimension and then the branch dimension. In a *reducing tee,* the branch dimension is smaller than the straight-through dimension, allowing the plumber to run smaller pipe off the main run. In a *regular-fitting tee,* all dimensions are the same, allowing the plumber to continue the same size piping.

BRASS FITTINGS

Brass fittings are used on steam systems and on potable water-supply lines. They can also be used on galvanized and cast-iron piping systems. Made in iron pipe sizes, brass fittings are available from ⅛ inch to 6 inches. These fittings have working pressures of 125 psi; extra-heavy fittings can withstand working pressures of 150 psi. The fittings are installed on the threaded pipe end and tightened using a pipe wrench. Several types are available. (See figure 2–13.)

Bushings. A bushing is used to connect pipes of different sizes. It is turned clockwise into a female fitting, reducing the size of the fitting.

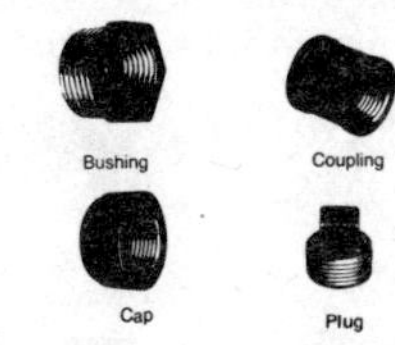

(Figure 2-13) Brass fittings

Caps. A cap is used to close a pipe that has a male thread on its end. The cap stops the flow of air, water, or other fluid. Caps come in a wide variety of sizes; they are simply screwed into place with a wrench.

Couplings. A coupling is used to join two or more lengths of pipe in a straight run. It has female threads on both ends and is turned clockwise onto the male threads of the pipe. Then two pipe wrenches (one at each end) are used to tighten the connection.

Plugs. A plug is a short length of metal with external (male) threads used to close the end of a pipe or fitting having female threads. A wrench is used to tighten the plug to prevent leaks.

BUSHINGS, I.P.S.

A bushing is a pipe fitting with both male and female threads and is used to connect pipes of different sizes. Most of the bushing screws into the larger pipe fitting, and then the smaller pipe screws into the bushing.

Bushings are commonly used, especially in repair and maintenance work. They are used on threaded pipe and plastic pipe. They should not, however, be used on natural gas systems; in fact, many local codes do no allow bushings to be used on natural gas systems.

(Figure 2-14a) Bushings

Bushings are made of various materials and in several types in iron pipe sizes. (See Figure 2–14a.)

Black Bushings. Black bushings are used in oil, heating, steam, and air lines.

Galvanized Bushings. These bushings are used in potable water-supply lines, in vent lines in drainage systems, and in air lines.

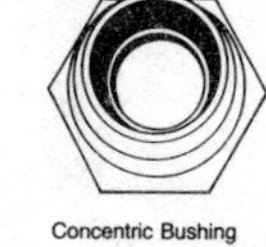

(Figure 2-14b) Eccentric and concentric bushing

Concentric Bushings. Concentric bushings, installed clockwise into a female fitting, enable a plumber to keep the center of the pipe as level as possible.

Eccentric Bushings. Eccentric bushings allow the draining of moisture or water that collects in the bottom of a pipe. (See Figure 2–14b.)

Flush Bushings. These bushings have threads on the exterior and interior walls and are used in close or tight pipe-fitting situations.

CAST-IRON SOIL PIPE FITTINGS

There are specific fittings for each type of cast-iron soil pipe, including beaded-end fittings, bell and spigot fittings, and no-hub fittings.

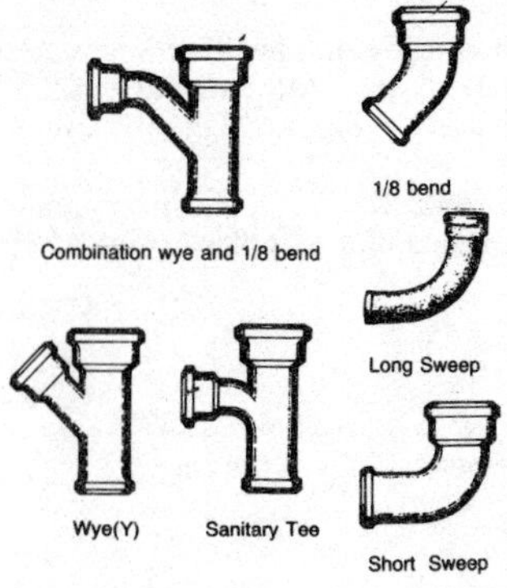

(Figure 2-15) Types of beaded-end cast-iron pipe

BEADED-END FITTINGS

Standard cast-iron beaded fittings can be used for 125 psi steam pressure or 175 psi water pressure. Fitting sizes range from 2-inch to 15-inch inside diameter.

Fittings for beaded-end cast-iron soil pipe come in two weights: service weight and extra-heavy weight. The fittings are installed with lead and oakum joints to make them airtight and watertight. The smooth inside areas of the fittings allow waste materials to flow through them easily. These fittings should not be used with gasket installations.

Following are descriptions of several types of beaded end fittings. (See Figure 2–15.)

Combination Wye and ⅛ Bends. A combination wye and ⅛ bend is used to accept waste water from fixtures. The fitting is usually placed into the wall as the last fitting in the line before the waste leaves the building and heads to the sewer or septic system. The ⅛ bend allows a gradual turn of the waste to prevent stoppages. (See Figure 2–16 for dimensions for this type of fitting.)

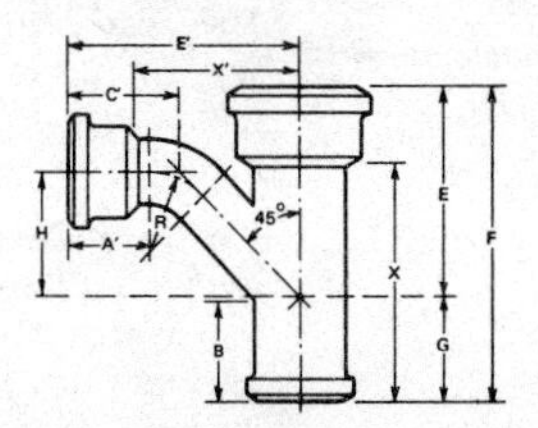

SINGLE COMBINATION WYE AND 1/8 BEND

Dimensions in inches[1]

Size (inches)	A'	B min.	C'	E	E'	F	G	H	R'	X	X'
2	2 3/4	3 1/2	4	6 1/2	7 3/8	10 1/2	4	3 3/8	3	8	4 7/8
3	3 1/4	4	4 11/16	8 1/4	9 3/4	13 1/4	5	5 1/16	3 1/2	10 1/2	7
4	3 1/2	4	5 3/16	9 3/4	12	15	5 1/4	6 13/16	4	12	9
5	3 1/2	4	5 3/8	11	14	16 1/2	5 1/2	8 5/8	4 1/2	13 1/2	11
6	3 1/2	4	5 9/16	12 1/4	15 7/8	18	5 3/4	10 5/16	5	15	12 7/8
8	4 1/8	5 1/2	6 5/8	15 5/16	20 1/2	23	7 11/16	13 7/8	6	19 1/2	17
3 × 2. .	3	4	4 1/4	7 9/16	8 1/4	11 3/4	4 3/16	4	3	9	5 3/4
4 × 2. .	3	4	4 1/4	8 5/16	8 3/4	12	3 11/16	4 1/2	3	9	6 1/4
4 × 3. .	3 1/4	4	4 11/16	9	10 1/4	13 1/2	4 1/2	5 9/16	3 1/2	10 1/2	7 1/2
5 × 2. .	3	4	4 1/4	8 5/8	9 1/4	12	3 3/8	5	3	9	6 3/4
5 × 3. .	3 1/4	4	4 11/16	9 1/2	10 3/4	13 1/2	4	6 1/16	3 1/2	10 1/2	8
5 × 4. .	3 1/2	4	5 3/16	10 1/4	12 1/2	15	4 3/4	7 5/16	4	12	9 1/2
6 × 2. .	3	4	4 1/4	9 5/16	9 3/4	12	2 11/16	5 1/2	3	9	7 1/4
6 × 3. .	3 1/4	4	4 11/16	10	11 1/4	13 1/2	3 1/2	6 9/16	3 1/2	10 1/2	8 1/2
6 × 4. .	3 1/2	4	5 3/16	10 3/4	13	15	4 1/4	7 13/16	4	12	10
6 × 5. .	3 1/2	4	5 3/8	11 7/16	14 1/2	16 1/2	5 1/16	9 1/8	4 1/2	13 1/2	11 1/2
8 × 2. .	3	5 1/2	4 1/4	10 7/8	10 3/4	14	3 1/8	6 1/2	3	10 1/2	8 1/4
8 × 3. .	3 1/4	5 1/2	4 11/16	11 9/16	12 1/4	15 1/2	3 15/16	7 9/16	3 1/2	12	9 1/2
8 × 4. .	3 1/2	5 1/2	5 3/16	12 1/4	14	17	4 3/4	8 13/16	4	13 1/2	11
8 × 5. .	3 1/2	5 1/2	5 3/8	13	15 1/2	18 1/2	5 1/2	10 1/8	4 1/2	15	12 1/2
8 × 6. .	3 1/2	5 1/2	5 9/16	13 11/16	16 7/8	20	6 5/16	11 1/4	5	16 1/2	13 7/8

(Figure 2-16) Cast-iron soil pipe fittings

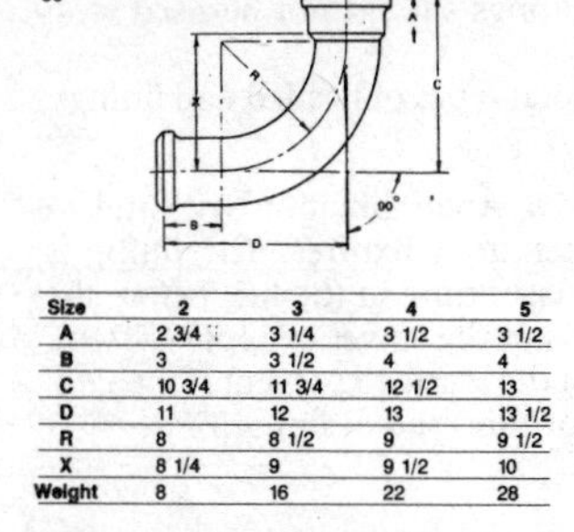

Size	2	3	4	5
A	2 3/4	3 1/4	3 1/2	3 1/2
B	3	3 1/2	4	4
C	10 3/4	11 3/4	12 1/2	13
D	11	12	13	13 1/2
R	8	8 1/2	9	9 1/2
X	8 1/4	9	9 1/2	10
Weight	8	16	22	28

(Figure 2-17) Fitting dimensions for a cast-iron beaded long sweep

Long Sweeps. A long sweep is used to make a vertical turn of the waste pipe. It can also be used on the horizontal run to accept and turn the flow of waste water. The long sweep helps to reduce the chance of a stoppage in the waste line. (See Figure 2–17.)

⅛ Bends. ⅛ bends are used to offset the waste or vent pipe. They can also be placed into wye fittings to make a combination wye and ⅛ bends. (See Figure 2–18.)

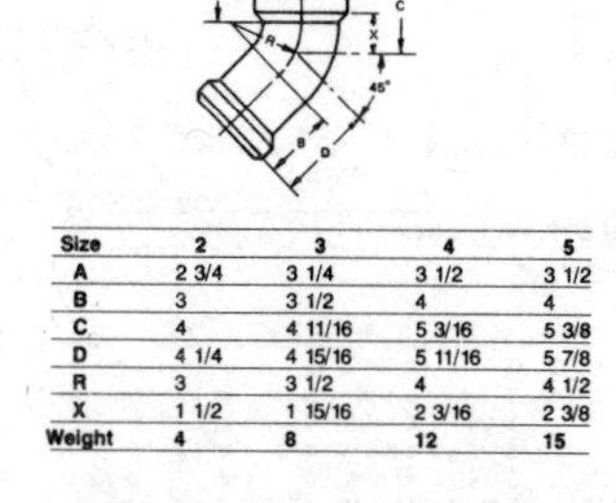

Size	2	3	4	5
A	2 3/4	3 1/4	3 1/2	3 1/2
B	3	3 1/2	4	4
C	4	4 11/16	5 3/16	5 3/8
D	4 1/4	4 15/16	5 11/16	5 7/8
R	3	3 1/2	4	4 1/2
X	1 1/2	1 15/16	2 3/16	2 3/8
Weight	4	8	12	15

(Figure 2-18) Fitting dimensions for ⅛ bend beaded cast-iron pipe

Sanitary Tees. Sanitary tees are placed into the piping system to accept waste water from kitchen sinks, basins, and vanity sinks. (See Figure 2–19.)

Short Sweeps. Short sweeps are usually placed at the end of a waste pipe. Generally, a short sweep should be used only to allow the passage of air (venting) or as an access point (with a cleanout) to relieve stoppages. (See Figure 2–20.)

Wye Fittings. These fittings allow different sizes of waste and vent pipes to enter the main run of piping. (See Figure 2–21.)

BELL AND SPIGOT FITTINGS

Cast-iron soil pipe with bell and spigot ends is used in waste and air venting systems. The fittings have one smooth or tapered end and one with a bell or hub. They come in sizes ranging from 2-inch to 15-inch inside diameter. They are available in two weights: service weight and extra-heavy weight. They can be installed using

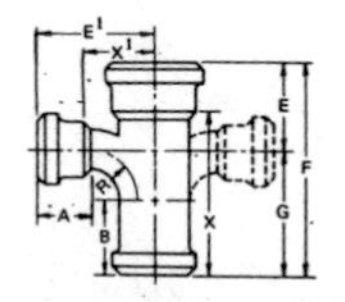

Single and Double Sanitary T-Branches

Size	A	B	E	E'	F	G	R	X	X'
2	2 3/4	3 3/4	4 1/4	5 1/4	10 1/2	6 1/4	2 1/2	8	2 3/4
3	3 1/4	4	5 1/4	6 3/4	12 3/4	7 1/2	3 1/2	10	4
4	3 1/2	4	6	7 1/2	14	8	4	11	4 1/2
5	3 1/2	4	6 1/2	8	15	8 1/2	4 1/2	12	5
3 × 2	3	4	4 3/4	6 1/2	11 3/4	7	3	9	4
4 × 2	3	4	5	7	12	7	3	9	4 1/2
4 × 3	3 1/4	4	5 1/2	7 1/4	13	7 1/2	3 1/2	10	4 1/2
5 × 2	3	4	5	7 1/2	12	7	3	9	5
5 × 3	3 1/4	4	5 1/2	7 3/4	13	7 1/2	3 1/2	10	5

(Figure 2-19) Fitting dimensions for single and double T-branches for cast-iron pipe

lead and oakum joints and neoprene compression joints. Gaskets make installation quick and easy.

Several commonly used bell and spigot end fittings are described below.

Cleanout Tees. Cleanouts are placed in the vertical and horizontal run of piping. The brass plug can be removed (counterclockwise) to help in case of stoppages. (See Figure 2–22a.)

Closet Bends. Closet bends are used in installing toilets. The bend is placed beneath the floor, a closet flange is placed over it, and the toilet bowl is tightened to the floor flange. Closet bends come in 3-inch and 4-inch sizes. (See Figure 2–22b.)

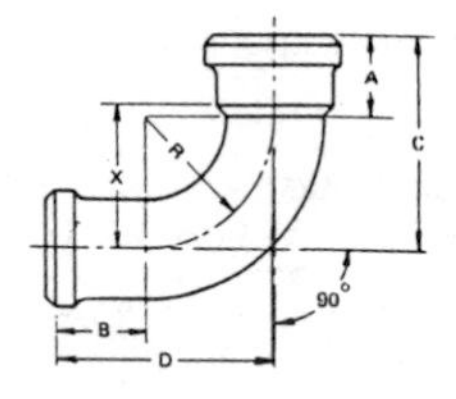

Size	2	3	4	5
A	2 3/4	3 1/4	3 1/2	3 1/2
B	3	3 1/2	4	4
C	7 3/4	8 3/4	9 1/2	10
D	8	9	10	10 1/2
R	5	5 1/2	6	6 1/2
X	5 1/4	6	6 1/2	7
Weight	6	13	18	23

(Figure 2-20) Short sweep fittings for beaded cast-iron pipe

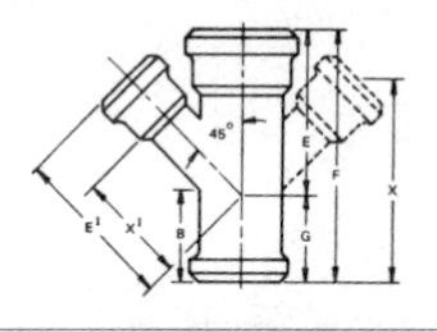

Single and Double Wye-Branches							
Size	B (Min)	E	E^1	F	G	X	X^1
2	3 1/2	6 1/2	6 1/2	10 1/2	4	8	4
3	4	8 1/4	8 1/4	13 1/4	5	10 1/2	5 1/2
4	4	9 3/4	9 3/4	15	5 1/4	12	6 3/4
5	4	11	11	16 1/2	5 1/2	13 1/2	8
3 × 2	4	7 9/16	7 1/2	11 3/4	4 3/16	9	5
4 × 2	4	8 3/8	8 1/4	12	3 5/8	9	5 3/4
4 × 3	4	9 1/16	9	13 1/2	4 7/16	10 1/2	6 1/4
5 × 2	4	8 7/8	9	12	3 1/8	9	6 1/2
5 × 3	4	9 5/8	9 3/4	13 1/2	3 7/8	10 1/2	7
5 × 4	4	10 5/16	10 1/2	15	4 11/16	12	7 1/2
6 × 2	4	9 7/16	9 3/4	12	2 9/16	9	7 1/4
6 × 3	4	10 1/8	10 1/2	13 1/2	3 3/8	10 1/2	7 3/4
6 × 4	4	10 13/16	11 1/4	15	4 3/16	12	8 1/4
6 × 5	4	11 9/16	11 3/4	16 1/2	4 15/16	13 1/2	8 3/4

(Figure 2-21) Wye fittings for beaded cast-iron pipe

Stack Base Fittings. These fittings are usually placed at the base of a stack, above the floor level. They have cleanout covers built into them for access to the waste pipe in case of stoppage. (See Figure 2–22c.)

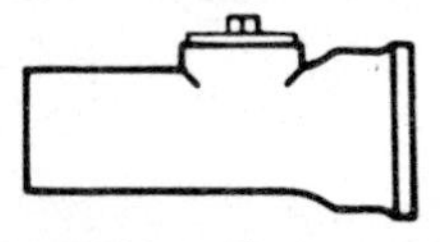

(Figure 2-22a) Cleanout tee for bell and spigot pipe

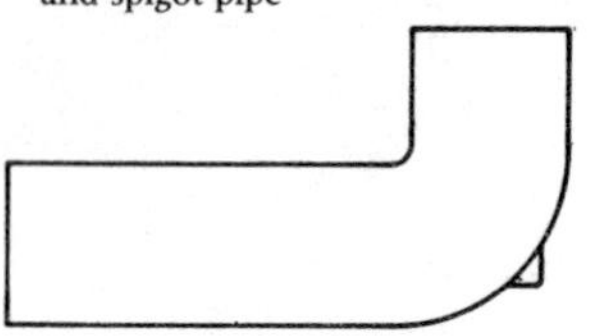

(Figure 2-22b) Closet bend for bell and spigot pipe

NO-HUB FITTINGS

No-hub fittings have neither a bell nor a hub at the ends. They have tapered ends and are manufactured in many more sizes than other cast-iron fittings—from 1½-inch to 10-inch, inside diameter.

All no-hub pipe and fittings are labelled with the manufacturer's name, the Cast Iron Soil

Pipe Institute's trademark, and the size. The label is about 1½ inches from the coupling joint.

No-hub fittings are easy to install and are common in many high-rise apartment buildings, hotels, and condominiums. They are selected and joined with a mechanical joint, which consists of a neoprene gasket, a stainless-steel outer jacket, and two stainless-steel clamps. The clamps are tightened using a torque wrench so that the pipe or fitting is not split or cracked. The clamps should be tightened alternately with about 60 foot-pounds of torque. (An attachment for an electric drill that allows two clamps to be tightened at the same time is available.)

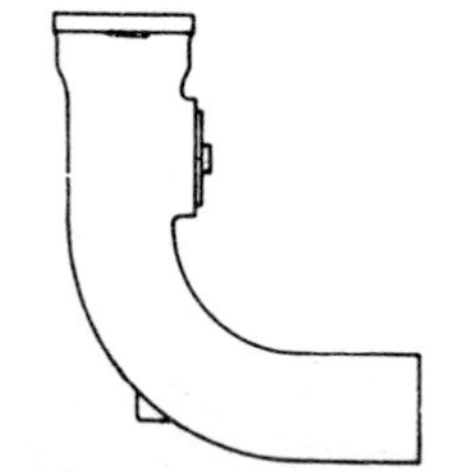

(Figure 2-22c) Stack base filling for bell and spigot pipe

Several commonly used types of no-hub fittings are described below.

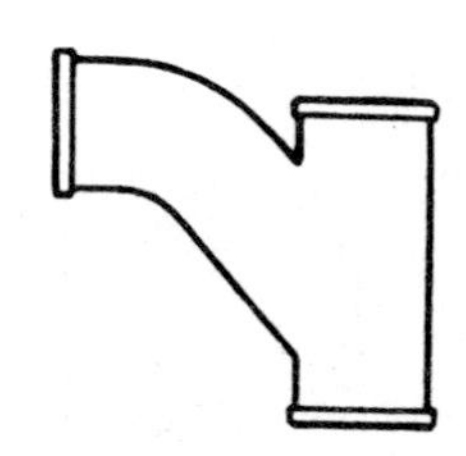

(Figure 2-23a) No-hub combination wye and ⅛ bend

Combination Wye and ⅛ Bends. These fittings are used to change the direction of the house drain and house sewer. They can be used vertically (⅛ bend up) or horizontally to accept discharged waste. (See Figure 2–23a.)

Long Sweeps. Long sweeps are used to make a long 90-degree turn, which helps to reduce the chance of stoppages. Long sweeps can be used to accept air or waste water in the vertical or horizontal position. (See Figure 2–23b.)

Sanitary Tees. Sanitary tees accept waste water from fixtures such as kitchen sinks, lavatories, and drinking fountains. They are usually installed in the vertical position. (See Figure 2–23c.)

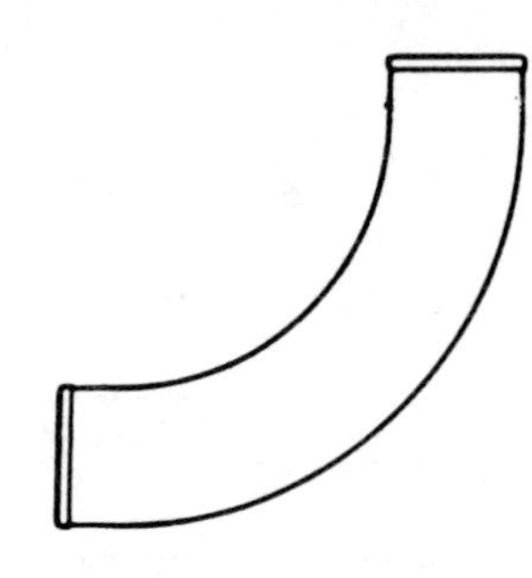

(Figure 2-23b) Long sweep

CLEANOUTS

A cleanout is a fitting with a removable plug, placed in a drainage pipe to access the piping system to relieve stoppages. Very important in any waste/drainage piping system, cleanouts provide an entry area for sewer snakes, flat snakes, sewer cleaning machines, and other cleaning devices.

In general, on pipes 4 inches or less in diameter, a cleanout should be the same size as the pipe it serves. For pipes over 4 inches in diameter, the cleanout should be at least 4 inches. Cleanouts must not be more than 75 feet apart, but should be installed every 50 feet and at every change of direction of a waste piping system. They must be installed to open either in the direction of flow of the drainage line or at a right angle to the flow.

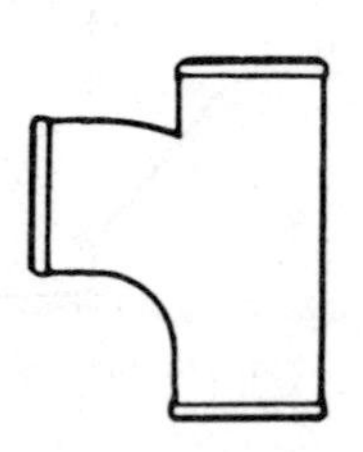

(Figure 2-23c) Sanitary tee

Cleanouts can be found in floors or walls, under a fixture, or in outside lawn areas. Cleanouts are installed in several ways, depending on the piping material. They should be soldered into copper fittings; cemented into ABS and PVC fittings; placed into cast-iron hubs with lead and oakum; and tightened onto no-hub fittings with a rubber and stainless-steel clamping system. Chrome or brass covers are usually placed over cleanout areas.

Cleanouts are made of different materials and are sometimes installed with various accessories. (See Figure 2–24.)

Cast-Iron Extended Cleanouts. Longer-length fittings, extended clean-outs are used to bring the cleanout itself to grade level.

Cast-Iron Regular Cleanouts. These fittings are used in basements and easily reached areas of the waste pipe.

Copper Cleanouts. These fittings are soldered into copper fittings for easy access to copper systems in case of stoppages.

Countersunk Cleanouts. With the plug set flush with the floor, countersunk-cleanouts are used to bring the cleanout to finished grade level.

Plastic ABS Cleanouts. Cemented into ABS fittings, these fittings are placed at every change of direction of the waste pipe.

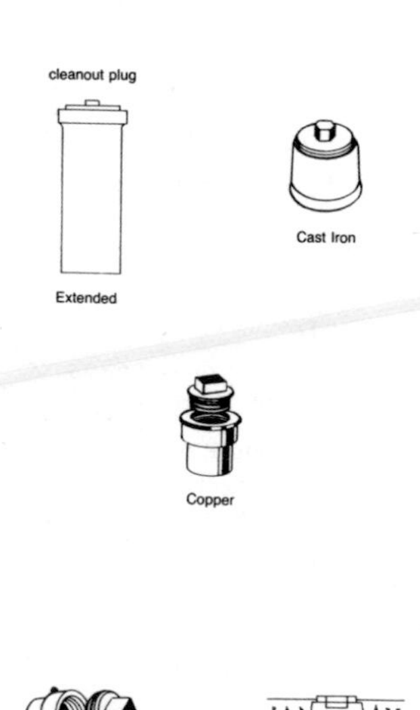

(Figure 2-24) Cleanouts

COPPER FITTINGS

Copper fittings are used on copper tubing in drainage, vent, and potable water systems. Each of the several types of copper fittings is designed for a particular purpose.

COPPER COMPRESSION FITTINGS

Copper compression fittings are used with fixture connections (hot and cold water) and on air lines, oil lines, ice cube water lines, etc. They are available in plumbing supply stores in a wide range of types and sizes.

These fittings require using two open-end or adjustable wrenches for tightening. Be sure not to overtighten because you may strip or break the nut used to tighten the ferrule.

(Figure 2-25a) Copper-to-copper compression fitting

Copper-to-Copper Compression Fittings. These fittings enable the plumber to solder a compression fitting onto copper tubing. Then, by inserting a flexible water-supply tube, a water line can be hooked up to the fixture. Gentle tightening of the compression nut will prevent leaks. (See Figure 2–25a.)

Double Compression Fittings. Double compression fittings are usually used to join two pieces of copper or plastic in a water or air line. You need only two wrenches to install this type of fitting. (See Figure 2–25b.)

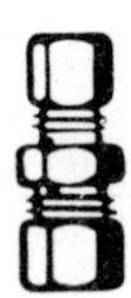

(Figure 2-25b) Double compression fitting

Iron-Pipe Compression Fittings. Iron-pipe compression fittings are used to change from a threaded (galvanized or brass) pipe to a flexible type of water line. The female fitting is attached to the threaded pipe, and the male fitting is tightened into the female fitting. Wrenches are then used to make the fitting watertight. (See Figure 2–25c.)

COPPER D.W.V. FITTINGS

Copper drainage fittings are used with copper pipe and are soldered together. They are available in a wide range of sizes from 1¼ inches to 8 inches. They are designed with a slight pitch to allow a smooth flow of sewage and air. The soldered joint area in a D.W.V. fitting is not as deep as it is in copper pressure fittings.

Copper D.W.V. fittings are manufactured in two ways: by casting or by a hammering (wrought) or drawn process. Cast fittings may

(Figure 2-25c) Iron pipe compression fitting

(Figure 2-25d) I.P.S. compression fitting

have sand holes or slight imperfections, but both types of fittings are acceptable for a drainage pipe system. The plumber should be careful, however, not to drop these fittings because they can be damaged easily, which then would cause a problem if they were to be placed on copper tubing.

DWV 90° Elbows

Size	"A"	"B"
1 1/4 –	1 3/16	1 3/16
1 1/2 –	1 7/16	1 7/16
2 –	1 15/16	1 15/16
3 –	2 15/16	2 15/16

(Figure 2-26) Copper D.W.V. 90° elbow

90-Degree Elbows. Cast or wrought 90-degree elbows are used to change the direction of flow of waste and air in a piping system. (See Figure 2–26.)

Street Elbows. Cast or wrought street elbows are used in tight places. They can also be used for "swing" joints or to make a trap. (See Figure 2–27.)

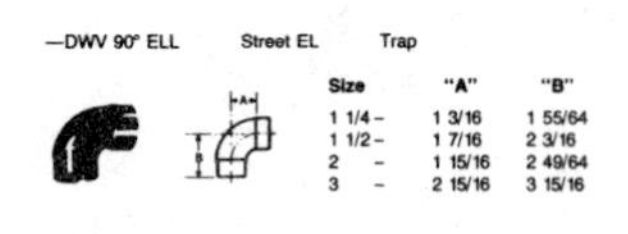

Size	"A"	"B"
1 1/4 –	1 3/16	1 55/64
1 1/2 –	1 7/16	2 3/16
2 –	1 15/16	2 49/64
3 –	2 15/16	3 15/16

(Figure 2-27) Copper D.W.V. street elbow

Traps with Cleanouts. Copper traps with cleanouts are used under showers and bathtubs and any other place that requires a trap. The cleanout enables the plumber to drain and remove local stoppages. (See Figure 2–28.)

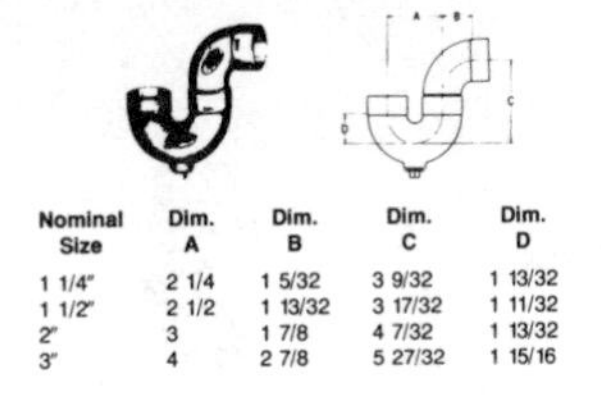

Nominal Size	Dim. A	Dim. B	Dim. C	Dim. D
1 1/4"	2 1/4	1 5/32	3 9/32	1 13/32
1 1/2"	2 1/2	1 13/32	3 17/32	1 11/32
2"	3	1 7/8	4 7/32	1 13/32
3"	4	2 7/8	5 27/32	1 15/16

(Figure 2-28) Copper D.W.V. traps

COPPER FLARED FITTINGS

Flared copper fittings are used on soft (annealed) copper tubing, primarily on underground water lines. They are available in ¼-inch to 3-inch sizes.

To flare the ends of copper tubing, insert a flaring tool into the tube. Then strike the tool with a hammer until a flare joint is completed. Use two wrenches to tighten the connection, but do not overtighten the fitting because you can strip or break it fairly easily. *Always be careful to protect the eyes when striking the flaring tool with the hammer.*

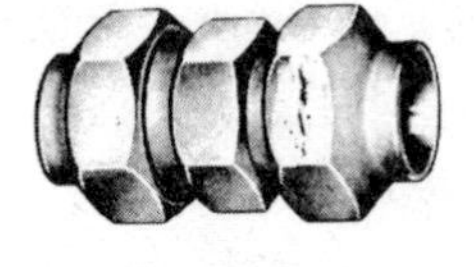

(Figure 2-29) Flared copper coupling

Couplings. A flared coupling is used to continue a run of copper tubing. It is one of the best ways to join an underground water line. (See Figure 2–29.)

(Figure 2-30a) Female flared copper adapter

Female Adapters. Female flared copper adapters are primarily used to tie an underground line to the curb stop, but they can also be useful in other piping systems. The female connection (iron-pipe-sized) is tightened onto any threaded pipe or nipple. (See Figure 2–30a.)

Male Adapters. Male flared copper adapters are also used to hook up to the city or local water supply. They are tightened into a female (I.P.S.) fitting. (See Figure 2–30b.)

Tee-Type Adapters. Tee-type flared adapters are used for multiple runs of copper tubing. The fitting and branch require a flared end on all tubing used. (See Figure 2–31.)

(Figure 2-30b) Male flared copper adapter

COPPER PRESSURE FITTINGS

Copper pressure fittings are used to supply potable water to plumbing fixtures. Water pressure in these lines is usually 40 to 80 psi. The fittings are available in sizes from ⅛ inch to 12 inches. They are soldered together for strength.

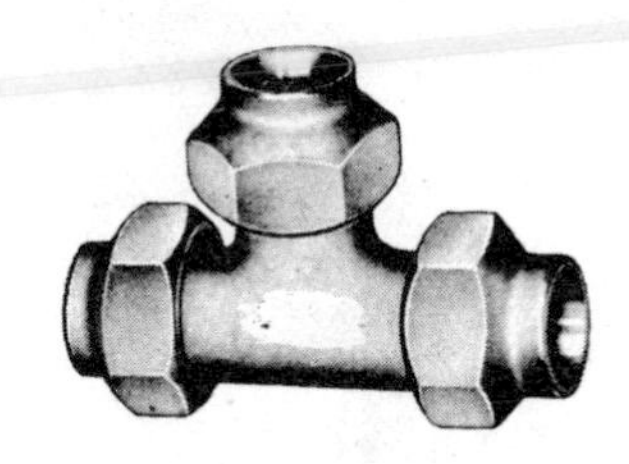

(Figure 2-31) Tee-type flared adapter

Copper pressure fittings are made by casting or by drawing or hammering (wrought). Cast fittings occasionally have holes in them, but either manufacturing method is acceptable. Pressure fittings have more surface area in the joint, making the joint stronger than a soldered joint in D.W.V. fittings, which have virtually no pressure on them except when the waste water runs. When installing these fittings, be sure not to overtighten them and to apply the proper heat so that the soldered joint remains strong. Do not overheat.

Drop (Ear) Ells. Drop ells are used for bathtub and shower diverters. Shower heads and tub spouts are turned into the female thread. The holes in the fitting (ear) allow the plumber to nail or screw the fitting into a piece of wood to prevent the water line from banging. (See Figure 2–32a.)

90-Degree Elbow Fittings. These pressure fittings are used to make 90-degree changes in the piping system. (See Figure 2–32b.)

(Figure 2-32a) Copper drop ell fitting

(Figure 2-32b) Copper 90° elbow fitting

(Figure 2-32c) Copper 90° fitting ell

(Figure 2-32d) Copper pressure tee

90—Degree Fitting Ells. These fittings, commonly called "street" ells, are used in tight spaces and to get around obstructions. (See Figure 2–32c.)

Tees. Tees are used to change direction from the main line of the water-supply system. They enable the plumber to run branch lines and air chambers for fixtures. (See Figure 2–32d.)

COUPLINGS

A coupling is a type of fitting used to extend or join lengths of pipe. Couplings are used on copper, steel, and plastic piping systems and, depending on the material, can be soldered, cemented, attached

(Figure 2-33a) Copper pressure coupling

(Figure 2-33b) CPVC coupling

(Figure 2-33c) Plastic (ABS) DWV coupling

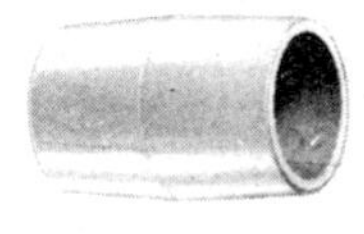

(Figure 2-33d) ABS repair coupling

with screws, or brazed. They are usually the same size as the pipes being connected; however, certain couplings can be used to reduce or increase the size of the pipe or tubing.

Copper Pressure Couplings. These couplings are used (by soldering) to extend or lengthen copper pipe or tubing. (See Figure 2–33a.)

CPVC Couplings. These fittings are used (by cementing) to lengthen chlorinated polyvinyl chloride (CPVC) pipe. Common sizes are ½ inch and ¾ inch. (See Figure 2–33b.)

Plastic D.W.V. (ABS) Couplings. These fittings are used (by cementing) to lengthen ABS waste pipe. Sizes range from 1¼ inches to 6 inches. (See Figure 2–33c.)

Plastic D.W.V. (ABS) Repair Couplings. With no stops, these fittings can easily slide up a pipe length to facilitate repairs. Sizes run from 1 ½ inches to 4 inches. (See Figure 2–33d.)

(Figure 2-33e) Reducing coupling

Reducing Couplings. Available in copper, steel, and plastic tubing sizes, reducing couplings are used to reduce or increase the size of pipe or tubing. (See Figure 2–33e.)

Steel Eccentric Couplings. These fittings are placed on the end of male threads or nipples to allow water (condensation) to drain out.

Steel-Pipe Couplings. Used with steel (black and galvanized) and brass pipes, these couplings allow the plumber to increase the length of the pipe run.

Cross T

Copper Pressure Cross Fitting

ABS D.W.V. Double Tee

Cast Iron Double Wye

(Figure 2-34) Cross fittings

CROSS FITTINGS

A cross fitting has four openings at right angles to one another. It allows the plumber to use the available space (often tight) to accept waste water or air (when the fitting is inverted) from two different directions. Actually taking the place of two separate fittings, a cross fitting cuts costs and labor and enables the plumber to design an efficient piping system that is easy to install and maintain.

Cross fittings are made in a variety of materials (copper, cast iron, galvanized steel, and plastic) and sizes. (See Figure 2–34.)

Cast-Iron Double-Wye Fittings. A double-wye fitting can accept waste piping from two different directions.

Copper D.W.V. Cross-Ty Fittings. These fittings are used to accept waste water in a copper system. If the ty is inverted, it can also accept air for venting purposes. Sizes run from 1¼ inches to 6 inches.

Copper Pressure Cross Fittings. Especially useful in tight spaces, copper pressure cross fittings allow the plumber to run potable water in various directions from the main feed line.

Galvanized Drainage Cross Long-Turn Ty Fittings. These fittings are used in galvanized drainage systems to accept waste water from two opposite directions. Sizes run from 1¼ inches to 12 inches.

Plastic D.W.V. (ABS) Cross-Ty Fittings. Used in plastic drainage systems, these fittings accept waste from two directions or, if inverted, accept air for venting purposes. Sizes range from 1¼ inches to 6 inches.

ELBOWS, DRAINAGE

A drainage elbow is a fitting that allows a run of pipe to change direction—usually 90 degrees or 45 degrees. It allows the plumber to run waste piping in standard walls. The fitting is designed with a pitch of ¼ inch per foot to accept and run waste water. Drainage elbows can also be used to carry air for venting purposes.

Several types of drainage elbows—of different materials and angle changes—are commonly used. (See Figure 2–35.)

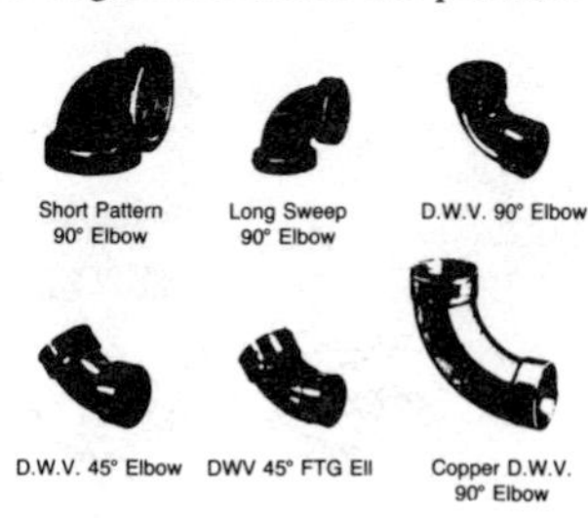

(Figure 2-35) Drainage elbows

Copper D.W.V. 90-Degree Elbows. These fittings are soldered to copper tubing. Sizes range from 1¼ inches to 12 inches.

Galvanized Drainage 90-Degree (Long-Sweep) Elbows. These fittings are used to carry waste water and air in drainage and venting systems. They are turned on standard I.P.S. threads and come in sizes from 1¼ inches to 12 inches.

Galvanized Drainage 90-Degree (Short-Sweep) Fittings. These elbows are used for venting (air) systems. They should *not* be used to carry waste water. The elbows are tightened on threaded pipe and come in sizes from 1¼ inches to 12 inches.

Plastic D.W.V. 45-Degree Elbows. These elbows are used to avoid obstructions and to make turns that will not cause a stoppage in the piping system. They are cemented onto plastic pipe. Sizes range from 1¼ inches to 6 inches.

Plastic D.W.V. 45-Degree Street Fittings. These elbows are used in tight places to make a long-turn 90-degree elbow. They are cemented into other fittings or onto plastic (ABS) pipe. The sizes available are from 1¼ to 6 inches.

Plastic D.W.V. 90-Degree Elbows. Designed to carry waste material out of the fixture area and into waste pipe, these plastic elbows are cemented onto ABS pipe. Sizes range from 1¼ inches to 6 inches.

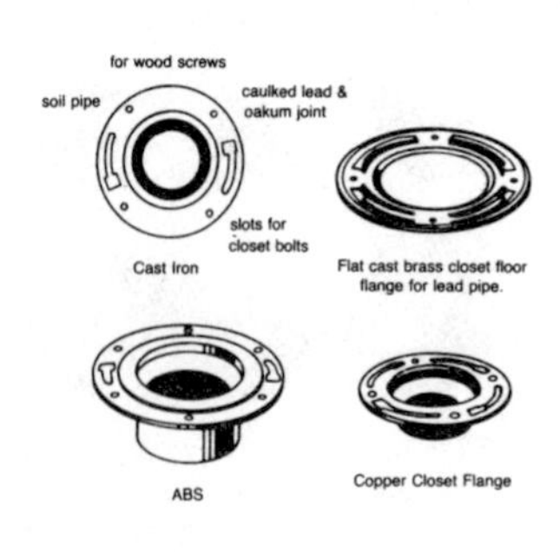

(Figure 2-36) Types of floor flanges

FLANGES, FLOOR

Floor flanges are used to secure a toilet bowl. They prevent the bowl from moving and, when used with a wax seal, maintain the water seal to the waste pipe.

Floor flanges are usually set even with the finished floor in a bathroom. Normal roughing dimension is 12 inches from the finished wall to the center of the floor flange. The flange, usually with an inside diameter of from 3 to 4 inches, should be screwed down to prevent the bowl from moving. A wax seal and toilet bolts are then used to make the water seal under the bowl. (See Figure 2–36.)

Improper installation of the floor flange can lead to waste-water problems and leaks that could damage the bathroom floor and the ceiling below.

ABS Floor Flanges. Cemented to ABS waste pipe, these floor flanges are screwed down; when it comes time to finish or "set" the toilet bowl, closet bolts are installed on the sides. The closet (toilet) bolts are gradually tightened, side to side. A wax seal is placed between the floor flange and the bottom of the toilet bowl. Closet flanges come in two sizes: 4-inch and 3-inch.

Cast-Brass Floor Flanges. These flanges are used with lead bends (4-inch and 3-inch) that are flared over the floor flange and soldered together.

Cast-Iron Floor Flanges. Used with cast-iron soil pipe, these flanges are installed using lead and oakum joints. The 4-inch size is the most common.

Copper Floor Flanges. Soldered onto copper tubing, these flanges come in two sizes: 3-inch and 4-inch.

FLARE-TYPE FITTINGS

Flare fittings are used on copper, steel, and plastic pipe. They range in size from ⅜ inch to 3 inches. Small flares are made with a flaring block, larger ones with a flaring tool. Larger flare fittings, ¾-inch and 1-inch, are used on underground water lines. A flare joint is one of the strongest ways to join two lines; properly made, a flare joint may even be taken apart for repair or maintenance work and then rejoined. Flare joints are used when soldering is not possible, and on copper tubing when a flame is not desirable (e.g., on an oil or gas line). (See Figure 2–37.)

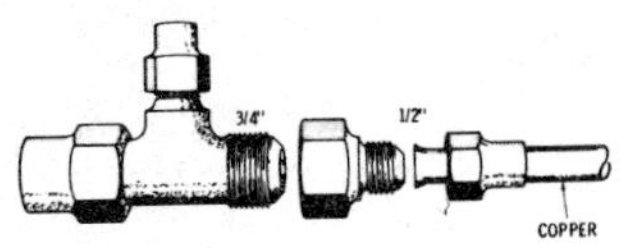

(Figure 2-37) Flare-type copper reducing fitting

Couplings. Flare-type couplings are used to join tubing of the same size.

Female-to-Flare Fittings. These fittings are used in a variety of ways, most often to change material in a piping system; for example, to go from a galvanized water line to copper tubing.

Flare Elbows, 90-Degree. These fittings are used to make turns. They prevent copper tubing from kinking or flattening out and thereby narrowing the tube.

Flat-Tee Reducing Fittings. These fittings allow branch lines to go from the main supply line to other size tubing.

Male-to-Flare Fittings. These fittings allow the plumber to change from one material to another. Available in ¾-inch and 1-inch sizes, they are commonly used on underground water lines to hook the house water service to the female curb cock left by the local water company in the street or at the property line.

Tee Fittings. Tee fittings enable the plumber to run branch lines from the main line.

FLASHINGS, ROOF

Roof flashings are used to make a roof watertight by preventing rain from entering the hole in the roof where the vent pipe passed through. It is extremely important to install flashings properly—before it rains.

Roof flashings come in many different sizes and are made of lead, copper, aluminum, or plastic. The flashing is slipped over the vent pipe and under the roof shingles. The corners are nailed to prevent the wind from driving rain under them. (Galvanized nails should be used.) Years ago a bead of tar was placed between the flashing's lead collar and the pipe. However, if the roof settles, the tar is useless and must be redone.

Most homes have D.W.V.-type roof flashings, but lead flashings are also available. (See Figure 2–38.)

D.W.V. Roof Flashings. Installed with many types of materials (copper, galvanized pipe, cast-iron, ABS, and PVC pipe), D.W.V. flashings come in many sizes, ranging from 1¼ inches to 4 inches. Larger sizes are used on commercial buildings and large apartment houses.

Lead Flashings. These flashings are made from flat lead. The chimney part of the flashing is made longer than the height of the vent pipe. The extra length is bent down inside the vent pipe to make it watertight. The flat part of the flange is usually bedded in with tar.

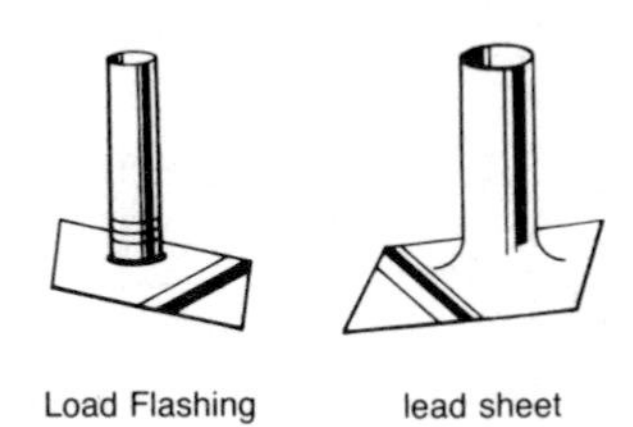

(Figure 2-38) Roof flashings

GASKETS, RUBBER

Rubber gaskets are used to join cast-iron soil pipe and fittings. They eliminate the need for lead and oakum joints. Insert the rubber gasket into the bell or hub area of the cast-iron pipe. Then coat the gasket and the spigot end of the pipe with a lubricant. Push, pull, or hit the fitting or pipe until it enters the gasket area. Remember: joints using gaskets must be well supported in vertical installations to prevent back pitch or bending.

Rubber gaskets of 2-inch, 3-inch, and 4-inch diameter are most commonly used, though sizes up to 15 inches are available from manufacturers. These gaskets are easily handled and stored. They can be reused and installed several times over. (See Figure 2–39.)

(Figure 2-39) Gaskets

Medium (SV) Silver Gaskets. These gaskets are used on medium cast-iron soil pipe and fittings.

Extra-Heavy (XH) Gold Gaskets. These are used on extra-heavy pipe and fittings.

HOUSE TRAP

A house, or building, trap is installed to prevent sewer gases from entering a home. Today it is usually installed inside a house, usually in the garage area with a concrete box built around it for easy access. House traps may also, however, be located underground outside the house, in which case some type of marker, such as a mushroom (venting) cap is brought above ground level. The trap must be installed so that there is access for repair and cleaning out of stoppages.

A house trap has two clean-out areas. The opening on the inlet, or hub, side is used for venting purposes. The trap is usually 4 inches in size and can be installed using lead and oakum joints or rubber gasket material. (See Figure 2–40.)

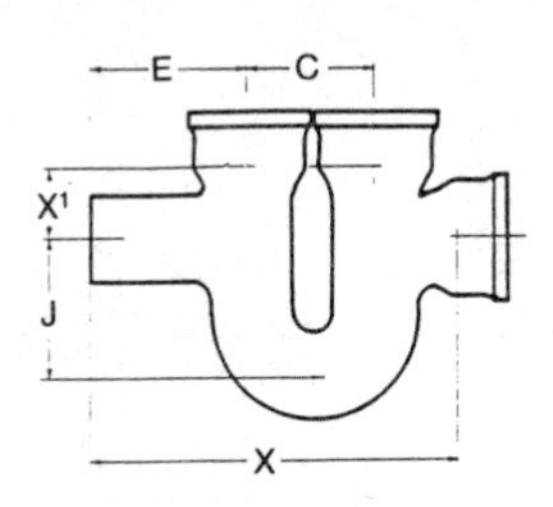

Running Trap w/Double Hub Vent

Size	X	X^1	C	E	J
3″-3″	15	2 1/2	5	7	5 1/2
4″-4″	17 1/2	3	6	8	6 1/2
6″-4″	21 1/2	4	8	9	8 1/2
6″-6″	21 1/2	4	8	9	8 1/2
8″-6″	27 1/2	5 1/4	10	12	11

(Figure 2-40) House trap

Not all plumbing codes permit house traps in the plumbing system, so be sure to check local codes.

MALLEABLE FITTINGS

Malleable fittings are available for black, brass, and galvanized pipe in sizes ranging from ⅛ inch to 6 inches; they can withstand working pressures of 150 psi. They are turned on the threaded end of a pipe and tightened with the use of two pipe wrenches.

Black fittings are used on air and natural gas lines. *Brass and galvanized malleable fittings* are used on potable water lines. *Galvanized fittings* can also be used on air lines and natural gas lines. (See Figure 2–41.)

Couplings. Malleable couplings are used to join pipes of the same size to lengthen the run.

45-Degree Elbow Fittings. These elbows are used to change

90-Degree Elbow

45-Degree Elbow

Coupling

90-Degree Street Elbow

45-Degree Street Elbow

Tee

(Figure 2-41) Malleable fittings

direction or to go around an obstruction. They restrict the flow of fluids less than 90-degree fittings.

45-Degree Street Elbow Fittings. These elbows are used in tight places to change direction.

90-Degree Elbow Fittings. Malleable 90-degree elbows are used to make right-angle changes in pipe direction.

90-Degree Reducing Elbows. These elbows are used to increase or decrease the size of the pipe and to change the direction of the pipe.

90-Degree Street Elbow Fittings. Street elbows are used in

Long Nipple

Close Nipple

Shoulder Nipple

(Figure 2-42) Nipples

tight places to change the direction of pipe and to make swing joints.

Tee, Reducing Fittings. Reducing tee fittings allow the plumber to reduce and change the direction of pipe without reducing the main supply line.

Tee, Regular Fittings. Regular fitting tees allow the plumber to branch off the main supply line with the same-size pipe.

NIPPLES

A nipple is a length of pipe, usually shorter than 12 inches, with male threads on both ends to join it to other fittings. There may be no unthreaded space between the ends (a close nipple), a small amount of unthreaded area (shoulder or short nipple), or a longer unthreaded area (long nipple). (See Figure 2–42.) Nipples have standard threads on each end and can be bought in standard pipe sizes from ⅛ inch to 12 inches. Since it is difficult to make nipples, especially small-sized nipples, without a special nipple chuck, it really does not pay a plumber to try to make nipples at the job site.

Nipples are available in brass, black, and galvanized. *Brass nipples* are used on potable water and steam installations. *Black nipples* are used on air systems, hot-water heating systems, and natural gas lines. *Galvanized nipples* are used on potable water lines, air lines, and natural gas lines.

The diameter, length, and material of a nipple is specified in a particular way: inside diameter first, then length, and then type of material. For example, a nipple specified as ½″ × 4″ galvanized steel is a 4-inch length of ½-inch diameter galvanized pipe.

PLASTIC FITTINGS

Following are descriptions of fittings designed for use with each of the major types of plastic piping.

ABS FITTINGS

ABS fittings are used on ABS drainage and venting systems. They are designed with internal pitch to allow waste to flow freely. Black in color, they are available in sizes from 1¼ inches to 6 inches; schedule #40 fittings are the most common. (See Figure 2–43a.)

ABS fittings are joined by solvent welding techniques. Be careful when aligning the fittings because once the "glue" has set, the pipes cannot be realigned.

(Figure 2-43a) ABS fittings

90-Degree Elbows. These fittings are designed to change the direction of the ABS waste or vent pipe.

45-Degree Elbows. These elbows are designed to avoid obstructions and to change the direction of pipe gradually.

Long-Turn Ty Fittings. These fittings help prevent stoppages; they accept waste and gradually change the direction of its flow.

Test Tees. These fittings are used to gain access to the waste system in case of stoppages.

Ty Fittings. These fittings are used to branch off the waste system and accept waste water or air for venting purposes.

Wyes. Wyes accept waste water or air.

CPVC FITTINGS

Chlorinated polyvinylchloride fittings are used on CPVC piping systems and are designed for higher pressures than are other types

(Figure 2-43b) CPVC fittings

of plastic fittings. In some states, CPVC fittings may be used on potable hot- and cold-water systems. They are also excellent for sprinkler systems and mist systems in greenhouses. They are available in ½-inch and ¾-inch sizes. They are joined with the appropriate CPVC solvent. (See Figure 2–43b.)

Bushings. CPVC bushings are used to reduce the inside diameter of the fittings; for example, from ¾-inch to ½-inch.

Caps. Caps are used to terminate the end of pipe to prevent leaks.

45-Degree Elbows. These elbows are used to change the direction of pipe gradually, creating less friction than a 90-degree elbow.

90-Degree Elbows. As with all 90-degree elbows, these CPVC fittings are used to change the direction of the piping system.

Street Fittings Elbow. These elbows are used in tight places.

Tees. Tees allow the plumber to run pipe in another direction off the main line.

PE FITTINGS

Polyethylene (PE) fittings are used with PE tubing or to change from another material, such as copper, to PE tubing. Designed for use at temperatures below 120°F, PE tubing and fixtures are used on field irrigation and cold potable water-supply systems. They are available in ¾-inch, 1-inch, 1½-inch, and 2-inch sizes. The fitting is inserted into the PE tubing, and then stainless-steel clamps are tightened over the tubing to make the joint watertight. Screwdrivers, torque wrenches, or nut drivers are used to tighten the clamps. Open flames should not be used near

(Figure 2-43c) PE fittings

these fittings, since they can catch fire or melt easily. (See Figure 2–43c.)

Couplings. PE couplings are used to join sections of PE tubing of the same diameter.

Male Adapters. Male adapters are used to screw into a female fitting.

90-Degree Elbows. These elbows are used to make tight turns.

Plastic-to-Female 90-Degree Elbows. These fittings are used to connect a standard nipple or threaded pipe to plastic tubing.

Tapped Tees. These special fittings allow the plumber to connect a nipple or pipe thread to a plastic line and then go in two directions.

Tees. Tees allow the plumber to branch lines off the main line.

PVC FITTINGS

Polyvinyl chloride (PVC) fittings are used on PVC piping, mainly in drainage, waste, and venting (D.W.V.) systems. They are also used in process piping of liquids and gases.

With a maximum service temperature of 140° to 150° F (depending on the type of piping and fitting), PVC fittings are usually white and designed with internal pitch to allow for the smooth flow of waste products. They are available in sizes from 1¼ inches to 6 inches.

PVC fittings used for drainage and venting are usually called schedule # 40 fittings. They are solvent welded to PVC pipe. (See Figure 2–43d.)

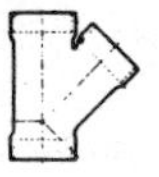
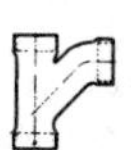
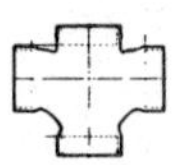

(Figure 2-43d) PVC fittings

Combination Wye and ⅛ Bends. These are used to gradually change the direction of waste pipe, without causing stoppages.

Double Sanitary Tees. Double sanitary tees allow the plumber to carry the vent pipe up and accept waste water from two directions. Inverted, it carries air up and out of the vent pipe system.

Wyes. Wyes accept waste and air from a direction different from that of the main line.

TRAPS

Traps are placed on a plumbing system to prevent sewer gases from entering a building. The water retained in the trap keeps odors from entering the room where the fixture is located. (See Figure 2–44.)

(Figure 2-44) Cast iron traps

Traps used in a home range from 1¼ inches to 2 inches; larger traps (2 to 4 inches) are usually made of cast iron and are found in commercial buildings or on sewer mains. Traps are necessary equipment, and their installation is usually governed by state and local codes. Be sure to follow code for the local area.

ABS Traps. These plastic traps are cemented with the proper cement for use on ABS systems.

Cast-Brass or Bath Traps. These traps are screwed on the threaded end of galvanized pipe or plastic or copper D.W.V. male adapters.

Cast-Iron (Durham) Traps. These traps are used on galvanized pipe and fittings.

Hub-Type Traps. Hub traps are used on cast-iron piping systems and installed with rubber gaskets or lead and oakum joints.

No-Hub Traps. No-hub traps are installed using compression gaskets and stainless-steel clamps.

P-Traps. These traps are used under fixtures. They are easily installed and removed for cleaning.

S-Traps. S-traps are common in many homes, particularly under kitchen sinks and basins. The trap is joined by a compression fitting into the waste pipe.

Sweat Traps. Sweat traps are soldered onto copper tubing. Unsoldering these traps can be dangerous because the flame often has to be very near the wall. Be careful not to start a fire inside the wall, and keep a fire extinguisher nearby at all times.

UNIONS

A union, usually placed near a fixture or piece of equipment, is a type of fitting that allows the plumber to break or go into the piping system without disturbing any fixtures or joints that precede or follow it in the piping line. Unions should be installed near any equipment thought to need replacement or repair.

A union has three parts: a collar and two mating parts joined by the collar. Unions are made in copper, brass, and galvanized iron; they are not made in cast iron. They can be soldered, cemented, or placed on threaded pipe.

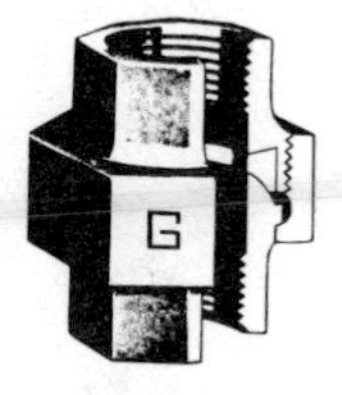
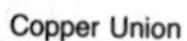

Copper Union

I.P.S. Union

(Figure 2-45) Unions

Copper unions are soldered near equipment on copper lines. *I.P.S. unions* are installed on galvanized, black, brass, and steam lines. (See Figure 2–45.)

FAUCETS AND VALVES

Faucets and valves are similar; in fact, a faucet can be considered a type of valve. Most faucets and valves come from the manufacturer with diagrams showing their component parts and with detailed installation instructions.

FAUCETS

A faucet is used to terminate a pipe run and is the final step in getting potable water to a fixture or other place of use. A spout is attached to many faucets to channel the outflow of water. Years

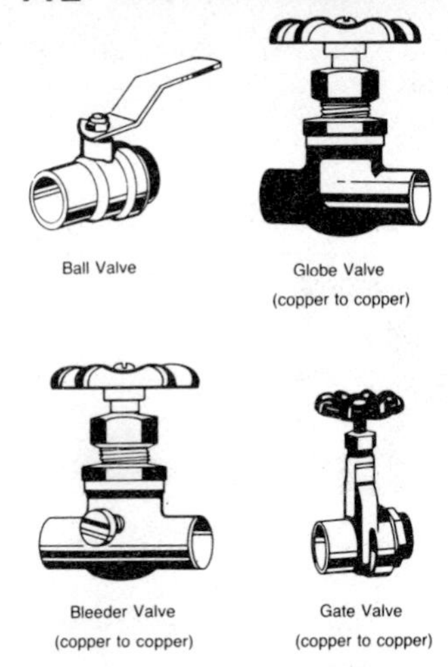

(Figure 2-46) Types of valve

ago faucets were made of brass and finished with chrome. Today brass, copper, plastic, and ceramic finishes are widely used as well as chrome, and sometimes gold.

Basin Faucets. Faucets for bathroom sinks usually have a fixed spout that does not swivel.

Kitchen Faucets. Kitchen faucets come with I.P.S., copper, and plastic connections. They can have individual valves (hot and cold) or a single handle. A spout is attached, which can usually swivel to allow water to be directed to a certain part of the sink.

Outside Faucets. Faucets installed outdoors—for example, on the outside of a house to allow the hookup of a garden hose—are usually built with an antisiphon device that prevents backflow. This anti-syphon device helps to protect the potable water from contamination.

VALVES

Valves are used to turn on and off parts of a piping system. They can be soldered, screwed on (I.P.S.), or cemented (with ABS and other plastic pipe). Proper placement of valves makes repair and maintenance work easy and quick. This placement is usually determined by code requirements, so be sure to check all local plumbing codes.

There are several types of valves. Some of the commonly used are discussed briefly here. (See Fig. 2–46.)

Backwater Valves. A backwater valve is a type of check valve installed to prevent a backflow of sewage from flooding a basement or lower level of a building. It is most often installed inside a building. If it is installed outside, plumbing codes may require that a pit but built around it to aid future maintenance. A backwater valve usually has a 4-inch inside diameter.

Ball Valves. Ball valves are made in all types of material, including plastic, and in many sizes. Only a ¼ turn of the handle is required to turn it "on" or "off." The disk found in most old-style valves has been replaced by a ball that quickly opens or closes the piping system. Ball valves are now replacing gate and globe valves.

Ballcock Valves. A ballcock valve is used in the toilet tank. When the water level in the tank drops, the valve opens to allow water to enter. When the water level rises, the float rises, shutting off the entry of water. Ballcock assemblies are made in plastic (replacing the old brass ones), which has made toilet tank repairs much easier.

Check Valves. A check valve is designed to prevent backflow in pipes. Water can flow readily in one direction, but any reversal of the flow causes the valve to close, stopping the backflow. Check valves are used in both horizontal and vertical positions. Horizontal check valves are placed in a level position on the pipe so that the valve disk inside falls back to center by gravity. A vertical check valve in a vertical piping system relies on a spring to help close it down.

Gate Valves. A gate valve uses a disk moving at a right angle to the flow of water to regulate the rate of flow. When a gate valve is fully open, there is no obstruction to the flow of water. In fact, when open a gate valve should always be used in the fully open position. Any other position can create a chattering inside the pipe and an annoying noise in the system. Gate valves are available in sizes from ¼ inch to 2 inches.

Globe Valves. A globe valve controls the flow of water with a compression disk. The disk, opened and closed by means of a stem, mates with a ground seat to stop water flow. Globe valves allow quick opening and closing of pipe supply lines. They are available in sizes from ¼ inch to 2 inches, sweat or I.P.S. They can be bought with or without "bleeder" openings. Bleeders, or caps, can be removed to release water or pressure on the system during repairs.

PART TWO

STANDARD PLUMBING PROCEDURES

3

Preventive Maintenance and Troubleshooting

With all types of plumbing installations, as with most equipment, proper maintenance and care can do much to prevent problems and to ensure the long life of the equipment. The plumber—and apprentice—should keep maintenance considerations in mind when installing pipe lines, fixtures, and appliances and should instruct customers on the proper use of the fixture or appliance. When problems do occur, the plumber should be able to diagnose the probable cause and understand how to remedy the situation. Such troubleshooting tests the plumber's ability to observe and understand how things work.

This chapter discusses preventive maintenance and troubleshooting of common plumbing installations. It is divided into several major sections: "Appliances," "Fixtures" (kitchen and bathroom), "Leaks," "Noises in the Piping System," "Pipe Stoppages," "Potable Water Systems," "Sewage-Handling Systems," and "Weather-Caused Problems." Each section discusses possible problems, how to prevent them, and how to correct them. Details concerning how to make specific repairs are included in Chapter 4, "The ABCs of the Trade," which presents a core of necessary information for a plumber.

APPLIANCES

There are several appliances that the plumber or apprentice may be called upon to install or repair. Among these are dishwashers, garbage disposal units, water heaters, and washing machines. Some require both hot- and cold-water lines; others only one or the other. Most require electricity, and some need waste and vent lines. All work must be done in accordance with local plumbing codes.

DISHWASHERS

A properly installed dishwasher is usually fairly troublefree. As with all appliances, however, there are a few guidelines to follow to ensure proper functioning. Some of these suggestions apply to the plumber or do-it-yourself installer; others are directed to the person using the appliance on a regular basis.

PREVENTIVE MAINTENANCE

- Be sure that a valve has been installed on the water-supply line. This allows the appliance to be repaired or replaced without shutting off the main water supply to the house.

- After the dishwasher's water, electric, and waste lines have been installed, check to see that the appliance is level. This will allow the pump to discharge the maximum amount of waste water.

- Make sure that the dishwasher is permanently secured (unless it is a portable unit on wheels). If the appliance is installed under a countertop, use a countertop mounting bracket and wood screws to screw the appliance into the base cabinets. This prevents the dishwasher from moving around during operation, thus decreasing noise and the chance of parts becoming loose or dislodged.

- In a house which has no heat during the winter (e.g., a summer house), drain the water line to the electric solenoid to prevent it from freezing during the cold weather.

TABLE 3–1

TROUBLESHOOTING A DISHWASHER

Problem	Probable Cause	Remedy
Little or no water pressure	Hot-water supply valve not open.	Check valve; turn the handle fully, counterclockwise, to ensure that the valve is open.
Noisy operation	Defective electric solenoid valve	Examine solenoid and replace it if found defective.
Will not start	Blown fuses Malfunctioning circuit breakers Defective or blown motor	Check fuses and circuit breakers. If they are not the cause of the problem, call appliance service or electrician. Motor may need to be replaced.

- Be sure to clear the dishes of food particles and then rinse them before placing them in the dishwasher. Large amounts of food and grease will eventually clog the discharge line.
- Save all manufacturer's instructions. They are very valuable when repairs or new parts are needed.

See Table 3–1 for basic troubleshooting of dishwasher problems.

GARBAGE DISPOSALS

Garbage disposers, or garbage disposals, as they are more commonly called, are electrical appliances installed under a kitchen sink where the duo (kitchen) strainer would normally be. After waste water and waste products are drained into it, a switch is turned on and the waste materials are ground up and washed through the kitchen trap into the waste-piping system. (See Figure 3–1.)

PREVENTIVE MAINTENANCE

Once the unit is installed, there is little the plumber or apprentice can do to maintain it to prevent trouble, other than to inform the user about its proper use and care:

(Figure 3-1) Cutaway of a garbage disposal

- Be careful not to let large objects, especially knives, forks, and spoons, fall into the drain area.
- Always use a lot of water when operating the garbage disposal. Water helps to carry the waste products through the drain pipes. Skimping on the amount of water used will in the long run create stoppages in the drainage system.
- Never leave the unit running for long periods of time. It is better to do two small loads than one large load.
- Read and save the manufacturer's use and safety instructions.

COMMON PROBLEMS AND HOW TO HANDLE THEM

The most common garbage disposal problem is a burned-out motor in the unit itself. The motor or entire unit then must be replaced. There are, however, a few other problems that may occur.

Unit Does Not Turn On. If the unit does not turn on, first check the switch to determine if it is working properly. This may require an electrician to check voltages at the switch and on the line supplying the unit. But remember, a burned-out motor is also a likely cause for failure of the machine to turn on.

Unit Turns On, But Waste Materials Do Not Drain from the Sink. If the unit turns on, but the waste water and waste products remain in the sink, there are several possible causes. Follow these steps to determine the cause:

Step 1. Turn off the electricity. *This is extremely important.*

Step 2. Disconnect the kitchen trap under the sink, and place a pail under the trap to catch waste products.
Step 3. Examine the trap to determine whether there are any obstructions.
Step 4. If obstructions are present in the trap, remove them and make sure the trap is free of objects and grease.
Step 5. Check to see if waste water has drained from the sink. If it has not, consider that there may be a problem with the disposal unit itself.
Step 6. Check that the electricity is turned off.
Step 7. Put your hand into the disposal to determine if any rags or other objects are lodged in it. Remove any obstructions.
Step 8. If water still does not drain through, consider that the problem may be a clogged drain line. These lines run through walls and ceilings.
Step 9. Clean the lines with an electric drain snake or waste wires.

Unit Turns On and Water Flows Out, But Waste Products Remain in Sink. If the unit turns on and waste water flows out, leaving solid waste products in the sink, the problem is probably with the blades in the disposal unit. It may be better to replace the entire unit than to attempt to repair it.

WATER HEATERS, ELECTRIC

An electric water heater heats water for kitchen and bathroom use. Cold water, supplied by the local water company or private well, is heated using 115 or 220 volts. The heating unit itself may have as small as a 6-gallon capacity and is usually easily installed, even in small or remote places. (See Figure 3–2.)

PREVENTIVE MAINTENANCE

Several routine practices will increase the life of an electric water heater and improve the quality of the water.

- Flush out the water heater once a year. Manufacturers recommend this practice to remove rust buildup in the bottom of the heater and prevent rust from entering the water lines.

- Maintain normal hot-water temperatures. Extremely high water temperatures over long periods may reduce the longevity of the heating unit.

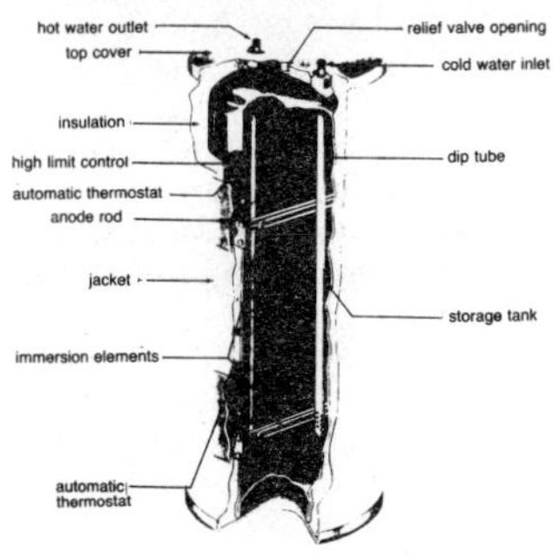

(Figure 3-2) Parts of an electric water heater

- Be sure that the proper F.H.A.-type relief valve is installed to protect the unit from exploding or rupturing.
- If a relief valve is found to be leaking, never plug it. Determine the cause of the problem.
- Follow the manufacturer's recommendations on maintaining the heating unit.

COMMON PROBLEMS AND HOW TO HANDLE THEM

There are several types of problems that may occur with electric water heaters.

No Hot Water. There are several possible causes for no hot water at all. First, check to see whether the circuit breaker or fuse is blown. If necessary, replace the fuse or reset the circuit breaker. If this does not correct the problem, you may need an electrician to check the voltage at the heater itself. It is possible that the heating elements themselves are defective or that the thermostats are malfunctioning. If the heating elements need to be replaced, be sure to drain the heating unit before beginning work. If, however, the thermostats need replacing, draining is not necessary.

Hot Water Turns Cold Quickly. If hot water turns to cold after only a few minutes of use, check upper thermostat and upper immersion element. The upper element is used to boost water temperature; if it is defective, the water will quickly turn cold.

Relief Valve Drips. If the relief valve drips, first check the temperature of the water coming out of the faucets. Match this temperature to the thermostat setting of the water heater. Then lower the thermostst setting and again check the water temperature. Any excessive hot water (steam) will make the relief valve go off and drip, thus protecting the heating unit itself. The valve is simply doing its job. If the water temperature, heating elements, and thermostats all seem to function properly and the valve still leaks, replace it. Use an F.H.A. pressure and temperature relief valve. This is required by many state and local plumbing codes. (See Figure 3–3.)

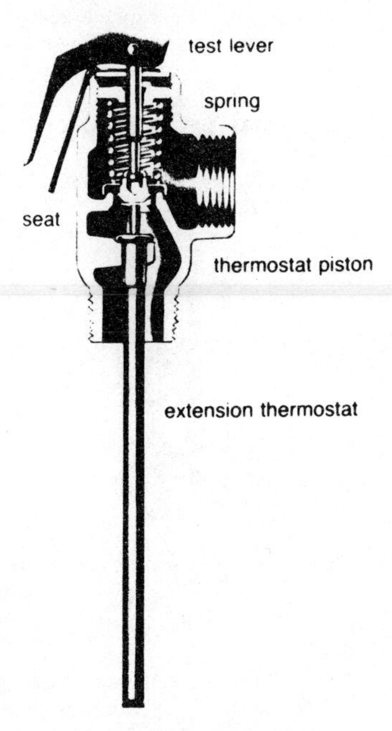

(Figure 3-3) Temperature and pressure relief valve

Rusty Water From Hot-Water Faucets. This is probably caused by a buildup of rust at the bottom of the heater. To prevent this problem, most manufacturers recommend that the water heater be flushed out once a year to remove any rust that is accumulating. If the heater is old, the amount of rust buildup may require that the unit be replaced; this depends, however, on the mineral content of the water in the area and the use of the unit. Old galvanized pipes may also need to be replaced.

Water Leaks from Inside the Tank. Once a water heater begins to leak from the storage tank itself, it should be replaced. Pin holes in the tank indicate that the inside walls have begun to rust out. This process weakens the tank walls, and eventually the tank itself could rupture and flood the basement or area where it is located.

A temporary repair of a leaking tank can be made: use neoprene washers with sheet-metal screws to stop the leak. However, remember that temporary repairs are just that—temporary.

WATER HEATERS, GAS-FIRED

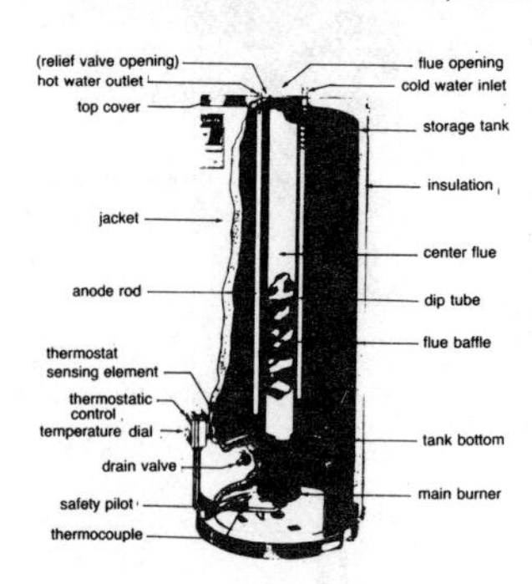

(Figure 3-4) Parts of a gas-fired water heater

A gas-fired water heater uses propane or natural gas to heat water supplied by the local water company or private well for domestic use. Most residential hot-water heaters come in 30- to 40-gallon sizes. (See Figure 3–4.) Gas-fired heaters must be installed according to local plumbing and fire codes and require a flue venting system.

PREVENTIVE MAINTENANCE

To increase the life of the unit and ensure proper functioning, several maintenance procedures should be followed:

- Flush the water heater occasionally to prevent rust buildup. (If rust should accumulate, it may mix with the water.)
- Maintain normal hot-water temperatures. Very high temperatures over a long period of time may reduce the longevity of the unit.
- Be sure that proper-sized gas lines and other installation equipment are used.
- Be sure that a proper-sized flue pipe and vent is used and that the venting system conforms to local codes.
- Be sure that the proper F.H.A. relief valve is installed. (See Figure 3–3.)
- If a relief valve is leaking, never plug it. Determine the cause of the leak, and correct it.

- Follow all other recommendations of the heating unit manufacturer.

COMMON PROBLEMS AND HOW TO HANDLE THEM

Although proper installation and maintenance will usually ensure that the gas-fired heater will operate properly for many years, some problems may occur.

No Hot Water. If there is no hot water at all, chances are good that there is no gas coming in or that the pilot light has gone out. Check to see whether the gas control valve is open and whether the pilot flame is lit. Then check the gas supply shut-off valve, and open it. Finally, check the main gas supply to the building.

Pilot Light Is Out. If the pilot light is out, turn the gas control knob to the pilot setting. Depress the pilot knob, and place a flame near the gas supply line thermocoupling (located next to the gas burner). Release when the pilot light remains lit. If the pilot light continues to go out, the thermocoupling is probably defective and should be replaced. In that case, follow the specific instructions of the manufacturer.

Excessive Hot Water (Steam). If there is too much steam, turn the thermostatic control (temperature setting) to a low setting, then check to see whether the water temperature decreases. If after adjusting this control, the water still remains hot, the entire heating control unit may need replacement. In that case, remember to drain the heater of water before starting work.

Rusty Water. Rusty water coming from the hot-water faucets is probably caused by rust buildup in the heater. To prevent this, flush and clean the heater once a year to prevent an accumulation of rust at the bottom of the unit. If rusting is severe, a new heater may be needed. (If galvanized pipes are old, they may also need to be replaced.)

Leak from Inside the Tank. A leak from the tank itself indicates that pin holes have developed in the tank wall, probably a result of inside rusting of the tank. Although the leak sometimes may be temporarily fixed with metal screws and neoprene washers, the repair (if possible) is only temporary. The heater should be replaced since the weakened walls could rupture and cause a flood.

TABLE 3–2
TROUBLESHOOTING WATER HEATERS

Problem	Probable Cause	Remedies
No hot water		
Electric	Blown fuses or defective circuit breaker	Replace or adjust,
	Defective heating element	Replace.
	Broken thermostat	Replace.
	Broken time clock	Replace.
Gas-Fired	No gas coming in	Check valves and adjust or replace.
	Defective thermocoupling	Replace.
	Dip tube installed incorrectly	Adjust—tube must be in the cold-water supply.
Relief valve drips	Excessive temperature setting	Lower setting.
	Defective valve	Replace.
Hot water turns cold quickly		
Electric	Defective upper or lower heating elements	Replace defective parts.
	Defective thermostat	Replace.
	Rust buildup	Drain and clean heater.
Gas-Fired	Dip tube in wrong inlet or broken	Insert tube in cold-water supply.
	Gas controller defective	Replace defective parts.

Gas Odor. If a gas odor develops, check fittings and valves for a possible leak. Use a soapy solution for testing. Also check flue-pipe connections, and look for holes in the pipe.

TABLE 3–2 (Continued)

TROUBLESHOOTING WATER HEATERS

Problem	Probable Cause	Remedies
Rusty hot water	Buildup of rust in heater	Drain and flush; if problem is severe, replace heater.
Water tank leaks	Rusting of inner tank walls causing pin holes	Replace heater (any repair is only temporary).
Gas smell	Hole in fittings, valves, or flue or poor connections	Replace any worn or defective parts; tighten connections.
Condensation	Heater installed in closed or confined area	Vent room or use louvers to allow air circulation.
Steam (extremely hot water) comes out of faucet		
Electric	Defective thermostat	Replace.
Gas-Fired	Defective gas control valve	Replace.
High operating costs	Sediment or lime in tank (rust)	Drain and flush (replace heater if rust buildup is severe).
	Water heater too small for job	Replace with larger heater.
	Wrong-size piping connections	Install correct-size piping.
	Lack of insulation on long runs of pipe	Insulate pipes.

WASHING MACHINES

Most washing machines are permanently installed in a basement or utility room. They require an electric line, hot- and cold-water lines, and a waste-pipe connection with a trap, and they must be properly vented according to local plumbing codes. Portable machines are simply rolled to the kitchen sink, and waste water is pumped into the sink.

PREVENTIVE MAINTENANCE

A properly installed washing machine is relatively troublefree. There are, however, some guidelines to follow to ensure long life.

- Be sure a three-pronged safety plug is used in connecting the machine to an electrical outlet.
- When leaving the house for a long period of time, always shut off the water-supply valves to the machine. The hoses freeze in cold weather if there is no heat, and they also can rupture if water pressure weakens the hose.
- Be sure that all manufacturer's instructions concerning the installation and use of the machine are carefully followed.

COMMON PROBLEMS AND HOW TO HANDLE THEM

No Water to Machine. If this occurs, first check that the water-supply valves are open. Also remove and clean the screens on water hoses to ensure that there is no obstruction to water flow. Then check the electric solenoid valve (may need special service person or electrician); if it is defective, replace it.

Water Pours Out from Top of Waste Pipe. If this occurs, move the plumbing waste-stand pipe higher. Also clean the trap, which may be clogged with soap particles.

Machine Does Not Cycle Properly. If the machine does not go through its normal cycles properly, the problem is most likely with the timing mechanism or within the unit itself. An appliance service person will probably handle this problem more efficiently than a plumber.

Water on Floor When Machine Discharges. When water appears on the floor while the machine is in operation, there is possibly a leak under the machine. A pump or pump hose may be defective or not properly connected.

FIXTURES

Plumbing fixtures are found in the kitchen, bathroom, and sometimes in the basement or utility room. Fixtures include kitchen sinks; lavatories (bathroom sinks); bathtubs; showers; and laundry trays (slop sinks). All of these fixtures are supplied with hot and cold water and accept and discharge waste into a waste-piping system. Although they differ in specific purpose and common use, the maintenance and proper uses of these fixtures, as well as the problems that may affect them, are similar.

The discussion below suggests ways to prevent problems with sink, tub, and shower drains and discusses common bathtub/shower and sink problems.

The toilet bowl, or water closet, as plumbers call it, is also a fixture. It is supplied with only cold water and, of course, accepts wastes that are discharged into the waste-piping system. Of all plumbing fixtures in the house, the toilet bowl has the most problems.

SINKS AND BATHTUB/SHOWERS

The basic ways to prevent problems, or to handle them once they occur, are similar with most types of drains—whether they be kitchen sink drains, lavatory drains, or bathtub/shower drains.

PREVENTING PROBLEMS WITH SINK, TUB, AND SHOWER DRAINS

- Keep all small objects (including small pieces of soap) away from drain area.
- Keep hair, granular soap, and all other foreign matter away from drain area.
- Do not allow grease to enter the kitchen drain. It will eventually clog the waste-pipe system.
- Run plenty of water to flush materials into the sewer system or septic system.

- If the tub drain has a plunger, remove it occasionally for cleaning.
- If shower drain has a plate over it, remove plate and any buildup on plate and on side walls of pipe. Use a flashlight to see small obstructions or buildup that can be easily removed—before they become a major problem.

COMMON BATHTUB AND SHOWER PROBLEMS AND HOW TO HANDLE THEM

Water Won't Stay in Tub. If this occurs, adjust the plunger of the waste and overflow equipment.

Water Won't Drain Out or Drains Out Too Slowly. This is the most common problem associated with tub and shower drains. In a tub, first remove the plunger and any hair or soap particles. If this does not solve the problem, the stoppage may be in the trap, which is usually the case with a shower. The trap is under the tub or shower, sometimes in a slab under a shower installation. This presents a problem, unless, as in some houses, there is an access door in an adjacent closet or in the ceiling below. If there is not easy access to the trap, you may need to cut an adjacent wall or the ceiling below to reach the trap. If the trap can be reached, take it apart and clean it thoroughly. To avoid this major work, most plumbers will try an electric or hand-driven sewer snake to clean the trap. Or in some cases, the plumber may pour chemicals (acids or alkalis, such as lye) down the drain in an attempt to dissolve the stoppage.

COMMON SINK PROBLEMS AND HOW TO HANDLE THEM

Several types of sinks are found in most homes: lavatories, or bathroom sinks; kitchen sinks; and laundry trays, or slop sinks. Although these sinks may differ somewhat, the problems that may occur are basically the same.

Pop-Up Valve Does Not Stay Up. Many bathroom sinks have a pop-up valve connected to the stopper. When the valve is lowered, the stopper is lowered and holds water in the basin; when the valve is raised, the stopper is raised and allows waste water to pass down through the drain. If the pop-up stem does not stay up, look under the basin and find the nut at the end of the rod extending into the tail-piece area. Tighten the nut—but do not overtighten it. When the nut is properly tightened, the pop-up mechanism should work properly.

Water Does Not Go Down Drain. This is the most common problem associated with all types of sinks. The causes are usually soap particles or hair in the case of lavatories, and grease and dropped utensils in the case of kitchen sinks.

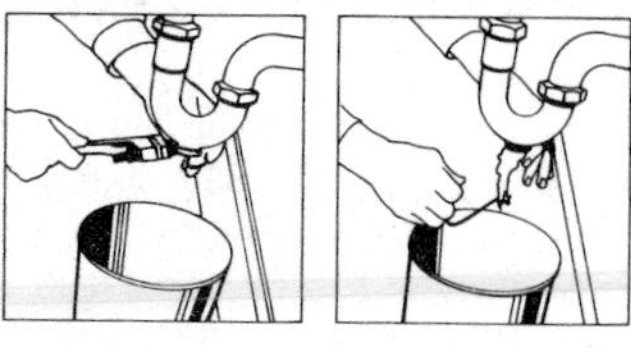

(Figure 3-5) Cleaning out a stoppage in a trap

If the sink has a pop-up valve, raise it and clean any accumulated hair or soap particles from the tail piece. Usually, however, the problem is in the trap. Check beneath the sink, and disconnect the trap. Be sure to place a pail under the trap area to catch any debris and waste water that may flow out as you disconnect the trap. (See Figure 3–5.) Clean the trap completely. If this does not solve the problem, the stoppage may be in the main drainage system in the house or in the sewer or septic system.

TOILETS (WATER CLOSETS)

Of all the fixtures in a house, the toilet bowl has the most problems. There are, however, some easy-to-follow rules to prevent problems, as well as some fairly easy steps to correct problems when they do occur.

PREVENTING PROBLEMS WITH THE TOILET

- Never allow any large object or rag to enter the toilet bowl. (Be careful that children do not drop even small toys down the bowl.)
- If you see an object lodged or floating in the toilet—for example, a small toy or piece of jewelry—remove it before flushing. Do not try to flush the object down.

- Check the water level in the tank periodically to make sure that it is the proper level.
- Don't stand on the toilet.

COMMON TOILET PROBLEMS AND HOW TO HANDLE THEM

Among common toilet bowl problems are leaks around the base of the bowl, constantly running water, and stoppages.

Leak Around Base of Toilet. If water appears beneath and around the toilet bowl, especially after flushing, the problem is most likely a defective wax seal. It is the wax seal that makes a watertight connection between the bowl and the floor flanges holding it to the floor. To correct this problem, remove the bowl and install a new wax seal. (While the bowl is removed, be sure to check the top of the soil pipe for any obstructions.)

Water Runs Constantly in the Tank. If water runs continuously, remove the top of the tank and check the water level and position of the ballcock valve and the float (or flush) ball.

If the water level is above the overflow pipe, the ballcock valve is not operating properly. Replace any worn parts, tighten any connections that are loose, or replace the entire ballcock assembly, if needed. Check the float ball to make sure no water has gotten into it. Replace it if necessary.

If the water is running constantly, but the water level is not above the overflow pipe, the problem is probably the float ball. Replace it.

Toilet Tank Sweats. This problem is caused by a difference in the temperature between the inside of the bowl and the bathroom air, and by humidity. It is a serious problem because, if not corrected, it can result in a rotted floor beneath the toilet. There are several possible solutions. Shelves may be placed under the tank to catch the sweat drippings. Or plastic liner, which acts as an insulator, may be used inside the tank to keep the porcelain of the tank from becoming too cold and to prevent water from condensing on the outside. Sometimes installing a dehumidifier in the bathroom will help. Covering the tank with a fabric, as some people do, helps little, if at all.

Toilet Tank Fills Up Too Slowly. If the tank does not fill up quickly enough, the problem is probably with the water supply.

Check to see that the supply valve is turned on fully—counterclockwise. Check to see whether the supply tube is kinked; if it is, straighten it out. Then check the ballcock system to see that it is working normally. If it is not, replace any worn or broken parts or the entire assembly, if necessary.

Toilet Stoppages. All human wastes and toilet paper (but not other types of paper) are soluble and will dissolve and pass through the toilet drain. If a stoppage occurs, it is because some object—such as a small toy—is inside the toilet bowl passageway. The object may become lodged and then paper and waste build up around it.

Use a closet auger to try to dislodge the object and clear the stoppage. A closet auger is a length of snake attached to a handle. Insert the auger into the bowl and turn the handle to guide the snake around the toilet drain area. You may be able to remove, break up, or snare the obstruction. (See Figure 3–6.)

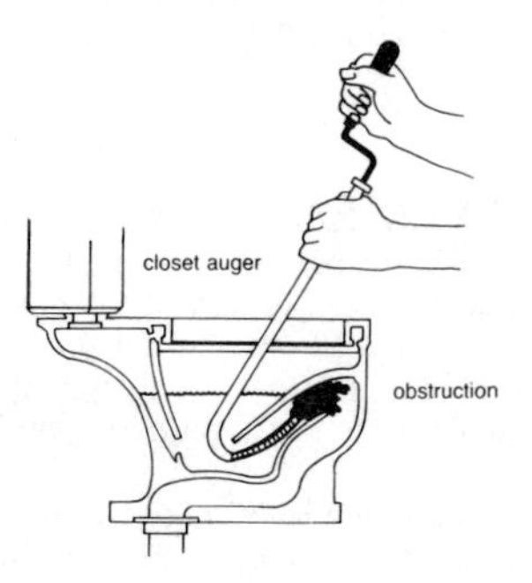

(Figure 3-6) Using a closet auger to relieve a toilet stoppage

If you cannot remove the stoppage with the closet auger, you may need to remove the toilet bowl, turn it upside down, and shake it a bit to try to remove the object. Sometimes a garden hose, inserted into the base of the toilet, will wash back any object lodged in the passageway. If this fails, the toilet bowl will need to be replaced.

LEAKS

Leaks are a common plumbing problem. They may involve appliances, fixtures, faucets, or pipes and fittings themselves. The leak may be immediately apparent—a dripping faucet, packing nut, or leaking drain pipe under the kitchen sink—or it may be difficult to locate, as in the case of a wet ceiling or wet wall.

APPLIANCE LEAKS

Appliance leaks, which usually occur during or immediately following the use of the appliance, may simply involve worn hoses or loose connections or may be more complicated, involving the operating mechanism of the appliance. See the section on "Appliances" earlier in this chapter for a discussion of common problems and how to handle them.

BASEMENT LEAKS

Water in the basement can come from many sources. Before trying to locate a leak in part of the piping system, make sure the water is not coming from another source. Heavy rainfall and high water tables often cause water to appear in the basement. So be sure the problem is, in fact, a plumbing problem before starting to do major work.

A plumbing leak in the basement may come from the waste, or drain, piping system or from the water-distribution system. A leak in the waste system can be in the main line leading out of the house to the cesspool, septic tank, or sewer, or it may be in any of the lines leading from the various fixtures and appliances in the house to the main line. A leak may appear only as a small spot and a slight odor that gets worse, but it should be repaired promptly because of the odor and possible health hazards.

A leak in the water-distribution lines can be in the main line coming into the house, in the line entering or leaving the hot-water heater, or in any line leading to an appliance or fixture. Water in the distribution lines is under pressure; a leak will only get worse, and therefore should be repaired promptly. Most distribution line repairs require shutting off the main valve controlling the water supply to the building. The exact nature of the repair depends on the type of piping involved, the extent and location of the leak, and other factors.

CEILING LEAKS

A ceiling leak can have many causes. Perhaps the most common cause is a leaking bathtub or shower trap. When tile can no longer keep water from penetrating the walls or floors, the wood or other subsurface material absorbs the water; eventually moisture builds up until a drip or more serious leak appears. Toilets that are improperly installed may also leak, resulting in a wet ceiling below. Another source of the problem could be a leak in the pipes and pipe fittings in the walls.

DISTRIBUTION LINE LEAKS

Leaks in the distribution lines—hot or cold—usually result in the loss of a considerable amount of water since the water in these lines is under pressure. The specific repair techniques to be used depend on the type of piping (e.g., copper, galvanized, or plastic), the size of the hole or crack, the location of the leak, and other factors.

DRAINAGE LINE LEAKS

Leaks in drainage pipes usually show up as small wet area, seepage, and possibly an odor that grows worse over time. The water in a drainage line is not under pressure, so there is seldom a large amount of water leaking. A leak in the drainage system may occur at the cleanout spot (often in the basement or garage), in the line leading out of the house, or in any of the pipes draining waste from the fixtures or appliances in the house. Again, the repair technique depends on the type of piping.

FAUCET LEAKS

A leak in a faucet is probably the most common problem encountered by a plumber. In older faucets, water is supplied by individual stems (hot and cold), and repair involves examining both stems and replacing all worn or defective parts. Modern faucets often are of the combination type, with many parts that can become loose, worn, or broken and cause a drip or more serious leak. The faucet usually has to be dismantled, the source of the leak identified, and the defective parts—if not the entire faucet assembly—replaced. Each manufacturer supplies diagrams of its faucets, showing all the component parts, and most plumbing supply stores carry all the parts needed for repairs. This emphasizes the importance of saving manufacturer's installation and repair instructions.

FIXTURE LEAKS

Leaks in bathtubs, sinks, and showers often involve dripping faucets or leaking traps. For bathtub leaks, be sure also to check waste and overflow equipment and shower diverters; for kitchen sinks, check sprays and strainers.

In addition to leaks at the bottom of a toilet bowl that may be caused by a defective wax seal, toilet leaks may also be due to worn fittings or defective float and ballcock mechanisms. Be sure to check all the parts in the tank as well as the cold-water line running into the toilet and the waste line draining the toilet.

PIPE AND JOINT LEAKS

Leaks at a joint, where two lengths of pipe are connected, are more common than leaks in the middle of a length of piping. When black, brass, or galvanized pipe or fittings leak, usually a part of the pipe and the joint need to be replaced. The leak may occur because of rust, overtightening that caused the threads to strip, unsecured vibrating pipes, or severe temperature changes. If simple techniques, such as tightening the threaded connection or using cement, do not work, the best solution is to replace a length of pipe, rethread the pipe ends, and make a new threaded joint.

When copper joints leak, they can usually be resoldered. The joint must be taken apart (unsoldered or cut out) and the fitting and pipe cleaned. Waste copper tubing is easier to repair than copper water line because waste piping has no pressure in it. Cold- and hot-water distribution lines—which have pressure—need to be drained to remove the pressure and volume of water before they can be repaired. This usually requires shutting off the nearby valves or the main valve on the line supplying water to the house. You cannot solder properly with water in the line because you will heat the water and cause steam to blow the fitting apart from the pipe or blow a hole in the solder joint.

When lead bends (elbows) leak, they should be replaced. A good example is the lead bend under the toilet bowl. Usually the bend thins out from usage over the years. This will cause water to leak out every time the toilet is flushed, and eventually the ceiling below will become wet and damaged. The lead bend should simply be replaced when it thins out.

When plastic fittings leak, clean and dry the area, then spread cement designed for use on plastic pipes and fittings over the area. If this does not stop the leak, a part of the pipe and joint may need to be replaced.

NOISES IN THE PIPING SYSTEM

There are several types of noises that can occur in a piping system. Some are simply annoying; others can indicate serious problems, such as a buildup of pressure in a pipe or vibrating and loose pipes.

BANGING NOISES

Banging noises in the piping system are often caused by improperly secured pipes that vibrate or hit against one another or another object. The first step is to locate the pipe or pipes causing the noise—usually in the basement. The pipe may need to be supported with an appropriate hanger or clamp.

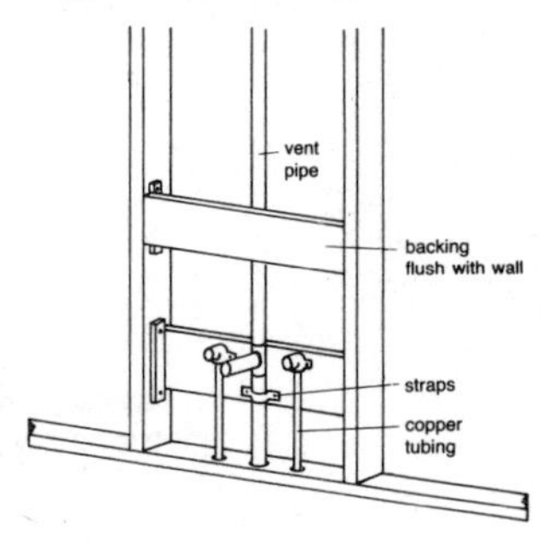

(Figure 3-7) Typical backing board setup behind a lavatory

Pipes placed in walls need backing. Usually wood is placed behind a fixture or pipe to allow a strap to be placed around the pipe and then nailed into the wood. (See Figure 3–7.) Normally, copper straps are used to hold copper tubing in place, while galvanized straps are used on galvanized and black piping

(Figure 3-8) Typical air chamber used to muffle a piping system

systems. Copper reznor hooks are used in basements on long lengths of copper tubing, while black reznor hooks are used on long runs of galvanized or black pipe. Perforated band iron (copper, black, galvanized, or plastic) is also used to support long runs of pipe and tubing, though it cannot be used on main drain lines.

SLAMMING NOISES

A slamming or banging noise that occurs when a faucet is opened or closed is usually caused by a lack of air chambers in the hot- and cold-water distribution lines. Air chambers, or air cushions, should be installed in back of each fixture to prevent noise created by air and water pressure. In fact, most state and local codes now require such air chambers, which keep piping systems stable and prolong the life of the equipment. (See Figure 3–8.)

OTHER NOISES

Other noises in a piping system can be caused by faucet washers that need replacing. The noise is usually a moaning or high-pitched sound that occurs only when that particular faucet is opened.

PIPE STOPPAGES

Stoppages may occur beyond the trap located below each appliance or fixture in the drain pipe that carries the waste to the main drain pipe of the house. They may also occur in the main line leading from the house to the sewer, cesspool, or septic tank. A blockage in a drainage pipe can often be cleared with a flat wire or snake. Use this equipment carefully, and be sure to clean it after each use. For severe pipe stoppages, a heavy-duty electric snake may be needed.

One type of stoppage is particularly troublesome—namely, a stoppage in a drain line caused by tree roots growing into the pipe. Chemicals may be used to kill the roots, but the only sure way to solve the problem is to use a large sewer-rooting machine that will completely clear the pipe and its connection with the septic tank or sewer. Licensed waste disposal companies have the type of equipment needed to clear this type of stoppage.

POTABLE WATER SYSTEMS

Safe drinking water is something most of us take for granted, and in most cases, this is justified. However, there are a few things that you can do to ensure clean, good-tasting water and to spot problems should they occur.

Perhaps the most common problem the homeowner may encounter—and call the plumber about—is discolored, usually rusty, water. This problem is usually caused by an accumulation of rust in the pipes or water heater. To avoid this, most water heater manufacturers recommend that the heater be drained and flushed once a year to eliminate any rust that may accumulate at the bottom of the unit. If rust accumulates, it may start to mix with the water, and rusty water will come out of the faucets. If there is severe rust buildup and the walls of the heater are rusting away, it is usually necessary to replace the heater. Old galvanized pipes may also need to be replaced.

Bad-smelling or bad-tasting water may also be a problem. This can be caused by the mineral content of the water or contamination of the water by chemicals. It can occur with local well water or with water pumped in from distant reservoirs. Contact the local water-supply company (city, town, or county) promptly. Explain the problem, and find out its cause. If you suspect that the water may not be safe, also contact the local health department.

TABLE 3–3
WHERE TO LOOK FOR CAUSES OF WATER-PRESSURE PROBLEMS

	Street water main	Curb stop	Water service	Branches	Valves	Stems, washers (hot and cold)	Aerator	Water meter
No water pressure at all	●	●	●					●
No water pressure at fixture					●	●	●	●
Low water pressure to fixture	●	●	●	●	●	●	●	●

If the water is not a health hazard, but is unpleasant tasting because of its hardness or mineral content, consider the use of water-conditioning or water-softening equipment. (See Figure 3–9.) It can usually be connected to the cold-water line coming into the house. The water passes through the device, which usually contains zeolite, a compound that absorbs the calcium and magnesium compounds that make water hard and often unpleasant tasting. The water then passes out of the purifier and continues its flow to the household faucets.

SEWAGE-HANDLING SYSTEMS

Sewage may be handled in several ways. It may drain from the house into a cesspool or septic tank with a leeching area or disposal field, or it may go directly into a municipal or county sewer system for treatment at a sewage plant. Cesspools and septic tanks depend on the action of bacteria to decompose the waste. (See Figure 3–10.) The waste water then drains out the cesspool into the surrounding soil or flows through pipes from a septic tank into a disposal field. (In many states it is illegal to install cesspools. Check local plumbing codes.)

Problems with sewage systems, especially with cesspool and septic tank systems, are common. The drain pipes in the house or leading out of the house to the pool or sewer may become clogged, causing sewage stoppages. The pools and tanks may also become clogged or filled so that sewage backs up through the pipes. These problems are usually handled by licensed waste-disposal companies. These companies, under the supervision of a plumber or health official, install cesspools (in areas where they are legal), septic systems, and sewer hookups, as well as handle the problems associated with the operation of these systems.

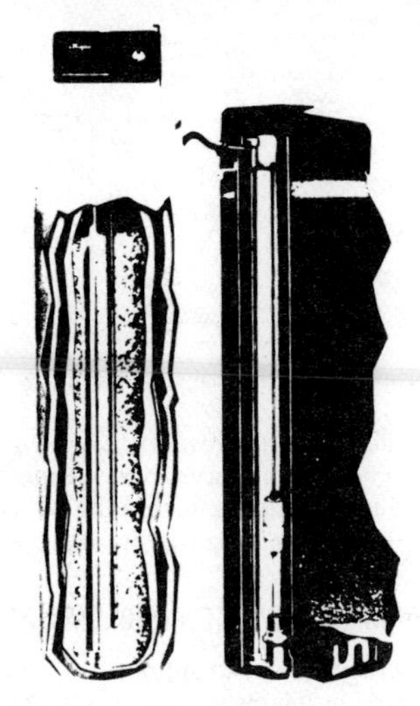

(Figure 3-9) Typical water softener

PREVENTIVE MAINTENANCE OF CESSPOOLS AND SEPTIC TANKS

There are several measures that the homeowner can take to decrease the chance of pipe stoppages and cesspool or septic tank overflow and backup. A responsible plumber should remind customers of these guidelines:

- Do not allow grease to enter the cesspool or septic system. Grease is not easily broken up by bacterial action and will build up and clog pipes.

- Do not allow too much water to flow into the pools or tank: it will interfere with bacterial action and fill up the tank. One

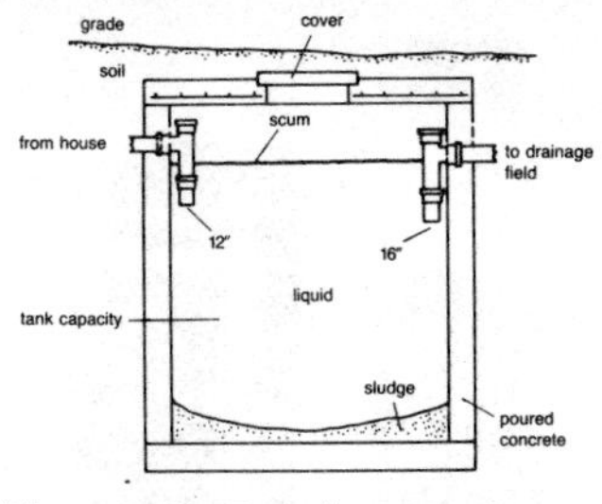

(Figure 3-10) Typical septic tank

way to limit the amount of water entering the tank is to hook the drain from the washing machine to a dry well (where legal) so that it does not enter the main house drain lines.

- Don't pour chemicals (such as paint thinners or photographic liquids) down the drain, and don't use chemical drain cleaners. Although drain cleaners may help break up a stoppage, the help is only temporary. These cleaners—and all chemicals—destroy the bacteria essential to the cesspool/septic tank operation, and thus in the long run are not beneficial.

COMMON SEWAGE PROBLEMS AND HOW TO HANDLE THEM

The plumber—and apprentice—should be able to diagnose sewage-system problems and inform the customer of the steps that should be taken. In most cases the services of a licensed waste-disposal company are required.

Wet Disposal Field. If the soil into which the cesspool or septic tank empties becomes too wet and soggy, another field must be provided. This requires installing a new cesspool in a different area, or in the case of a septic tank and leeching area, digging up the field and repositioning or replacing the system so it drains in another area.

Backup of Sewage. Backup of sewage can occur because the drain pipes from the house to the tank are clogged or because the cesspool or tank is filled to capacity. If the lines are clogged, electrical snaking will probably be necessary. If the pool or tank is full, it must be emptied. In either case, the services of a licensed waste-disposal company are probably needed.

WEATHER-CAUSED PROBLEMS

The problem of frozen pipes is quite common every year. The plumber must be familiar not only with how to remedy the problem, but also with what can be done to avoid the problem or at least minimize the chance of its occurring again.

Galvanized and brass water lines located underground, if they are installed properly, should not freeze. They are strong pipes that will not burst easily. Plastic and copper pipe (both rigid and flexible) have a little "give" and usually will not burst. However, there may be a problem with rigid fittings used with plastic and copper pipe: they may crack or burst under very cold conditions. Avoid placing fittings underground if you can, because once the ground freezes it is almost impossible to repair them if they leak.

PREVENTIVE MAINTENANCE

There are several steps that can be taken to minimize the chance of frozen pipes, and the plumber or apprentice should recommend them to homeowners. First, try to insulate all pipes, especially pipes in uninsulated exterior walls. If extremely cold weather is expected or problems with the heating system cause a house to become very cold, let a little water run from the faucets. This will keep the water moving in the pipes and should prevent freezing. An electric light bulb can also be positioned near—but not touching—a bare pipe. Heat tapes, properly installed, can also prevent frozen pipes, but check local fire codes to determine if the use of such tapes is permitted.

Most recommendations for preventing frozen pipes apply to unoccupied houses, expecially summer houses left unoccupied for the winter months.

PREVENTING FROZEN PIPES IN AN UNOCCUPIED HOUSE

- Try to have the main water supply to the building shut off. This may involve shutting off the water supply at the curb or at the main valve where the water line enters the house. For houses supplied by an individual well, the valves should be opened and the water allowed to drain back into the well.

- Open all valves and drain all pipes.
- Drain radiators, boilers, and hot-water tanks.
- Flush toilets.
- Add antifreeze to all traps that cannot be drained, including the water closet.

UNFREEZING PIPES

If brass or galvanized underground water lines become frozen, the plumber can attach electrical terminals from a welding machine to thaw the pipes. Underground copper pipes can sometimes be unfrozen the same way. However, underground plastic pipes cannot be thawed using electricity, since plastic does not conduct electricity. To thaw plastic pipes, try to gain access to the area of the frozen pipe and apply hot water (not too hot) or hot towels.

Frozen distribution lines in the house can be thawed using an electric blanket or hot towels wrapped around the pipe, hair dryers, drop light, or a torch (use carefully!). Do not heat only one small area, because then steam can build up and the pipe can blow or rupture. Also, be sure to start heating at the valve or faucet end of the pipe—again to prevent the possible buildup of steam. Open a faucet or valve to relieve pressure on the system.

4
ABCs of the Trade

The jobs that a plumber or apprentice may be called upon to do are many and varied. This chapter summarizes basic plumbing know-how.

First, the basic piping systems that may be found in a building are outlined. Then techniques for cutting and joining pipe are explained. These skills are used in virtually all plumbing work. Guidelines for installing and repairing different types of appliances, faucets, and fixtures are also given. A brief discussion of potable water systems is included and, finally, a discussion of roughing in, the first stage in plumbing installation, completes the chapter.

Wherever possible, instructions are given in an easy-to-follow, step-by-step approach.

BASIC PLUMBING SYSTEMS

There are several separate piping systems in a house or other building. These include the cold-water system, the hot-water system, the drainage system, the heating system, and, in some cases, natural-gas piping and other systems.

COLD-WATER SYSTEM

The potable cold-water system includes the water supplied to the house or other building by the local water-supply company or from a well. The water comes through a large pipe called a *main* that runs through the street or connects to the well. Mains are run underground and must be installed below the frost line to prevent freezing during winter months.

A *water-service line* runs from the main to the building. In a typical water-service line, there are at least two valves: one that is usually near the curb or property line; the other just inside the

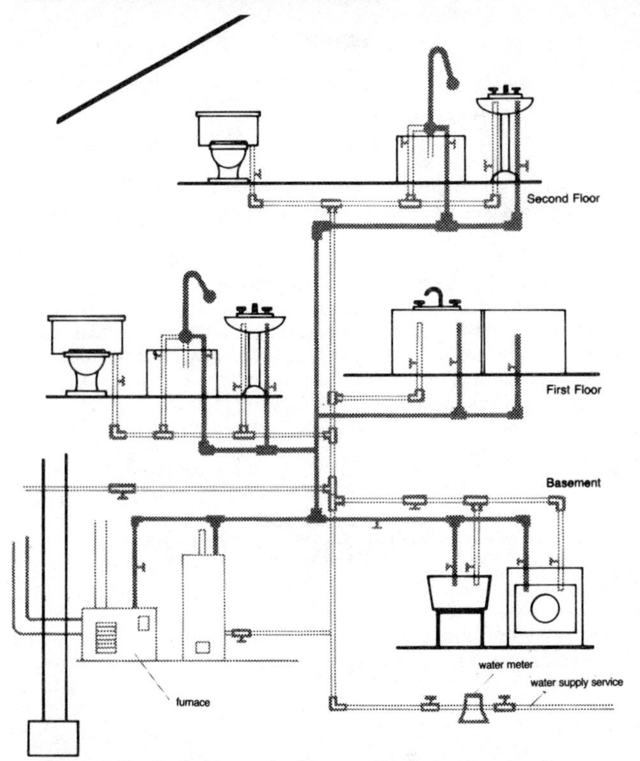

(Figure 4-1) Typical hot and cold water distribution lines in a house

building where the line enters. Closing the valves shuts off the flow of water into the building, permitting repairs.

After the point where the water-service line enters the building, a maze of *distribution lines* carry the cold potable water to various fixtures and appliances in the house. One line also supplies water to the hot-water heater to provide hot water for domestic purposes.

Most distribution lines also have valves so that water flow to a particular fixture or appliance can be shut off to allow repairs or other work. (See Figure 4–1.)

WATER METER

A water meter measures the amount of water consumed in a building. Many years ago water meters were often located in basements or crawl spaces, requiring the meter reader to enter the building to take a reading. Today, many water meters are found outside the building, at the curb or property line. The meter measures the amount of water being supplied to the building in gallons or cubic feet. The readings are usually taken periodically during the year (often quarterly) and are used to calculate the water bill for the building. (In some areas the amount of water supplied is also used to calculate sewer bills.)

HOT-WATER SYSTEM

Hot water needed in kitchens, bathrooms, utility rooms, and other places for various fixtures and appliances is known as domestic hot water. It should not be confused with the hot water in a heating system; the piping for the two systems is never connected.

Cold water supplied to the building can be heated in a special water heater (either electric or gas-fired) or in the same furnace that is used to heat the building. In all cases, the cold water is supplied to the heating unit through one pipe or tube and leaves the unit through another line. The hot water then flows through a maze of distribution lines to the various fixtures and appliances that require hot water, such as sinks, dishwashers, and washing machines (but not toilets).

The temperature of the hot water is regulated by controls on the heating unit. In general, it is best to avoid setting this temperature too high. Excessively hot, steamy water can damage heating units, valves, and piping systems. Temperature and pressure relief valves, installed on all water heaters, will go off to prevent the heating unit from exploding or rupturing.

DRAINAGE SYSTEM

The drainage system carries all waste material away from the fixture or appliance and out of the building. The system consists of pipes to carry the waste and pipes used for venting. The waste material is carried to a cesspool or septic tank or through sewers to a sewage-treatment facility.

Waste materials release gases that are odorous and sometimes poisonous or even explosive (for example, methane). To prevent gases (and odors) from entering the building, traps are installed under all fixtures and appliances. The water in the trap prevents odors from rising. Each trap is connected to a vent pipe leading through the roof. Vent pipes carry odors out into the air, relieve pressure in the system, prevent any siphoning effect that might occur from emptying a trap and, by allowing fresh air to circulate in the system, prevent slime and corrosion.

Waste pipes, traps, and vent pipes combine to enable the drainage system to function properly. (See Figure 4–2.)

HOT-WATER HEATING SYSTEM

In a hot-water heating system, hot water is used to heat a building. The cold potable water entering a building is fed into a hydronic boiler. A fuel, such as natural gas or oil, is used to heat the water, usually to a temperature of about 180°F. The hot water is then conducted, usually through copper tubing, to the radiators or baseboard radiating units in the various rooms of the house.

The hot water in the copper tubing heats up aluminum fins in the radiating units. This, in turn, heats up the air surrounding the unit. The hot air circulates in the room and rises to the ceiling. It then cools off, drops back to floor level, and is recirculated under the baseboard or radiator to be reheated around the aluminum fins. Meanwhile, the water moves back to the boiler, is reheated, and returns to the radiating units.

A thermostat installed in one or several rooms of the building controls the amount of heat provided. It is set at a specific temperature; when the temperature falls below that level, the boiler turns on, and heated water is sent through the system. Thermostats should be positioned carefully to accurately reflect ambient temperatures. If located too close to a door or window or in a draft, the thermostat will not provide an accurate reading of the room temperature.

There are several other ways to ensure a well-functioning hot-water heating system. For example, thick rugs should not be

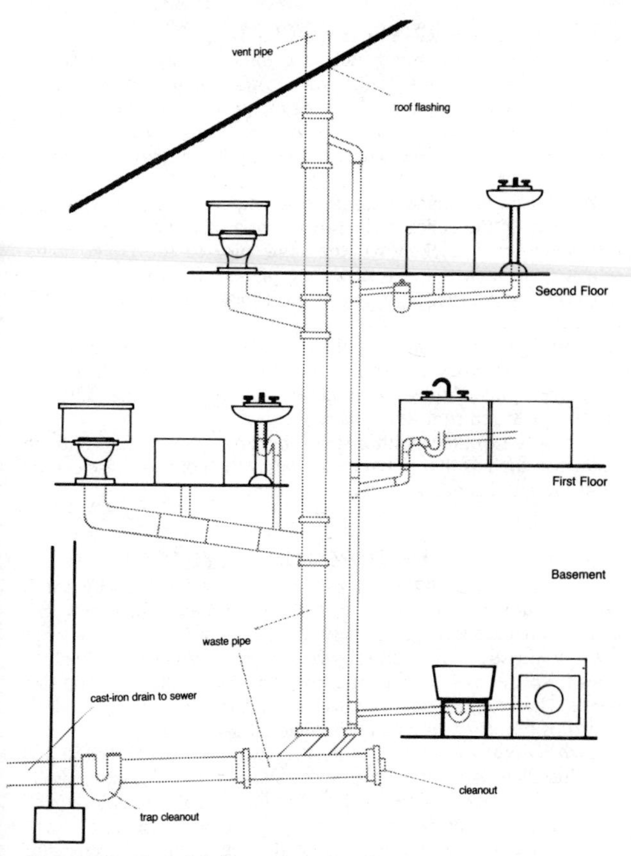

(Figure 4-2) Typical drainage system in a house

installed around the bottom of baseboard radiators (leave a two-inch clearance), nor should furniture be positioned near the radiating unit, blocking the recirculation of air.

STEAM HEATING SYSTEM

Steam heating systems are not as common today as they once were, but they are still found in some houses, apartment buildings, and commercial buildings. In a steam heating system, water is heated in a boiler to the boiling point; the steam is then carried to radiators in the building, where it gives up its heat and condenses back to water.

Most steam heating systems have one pipe. Steam flows through the pipe from the boiler to the radiator and then the condensate—the water—runs back down the same pipe to the boiler to be reheated. In a two-pipe system, one pipe supplies steam to each radiator, and another pipe returns the condensate to the boiler for reheating.

Steam lines are usually installed using black steel pipe and cast fittings. These pipes and fittings must be carefully installed with the correct pitch if they are to allow steam to rise to the radiators and water to return to the boiler.

One of the troubles with a steam heating system is the noise it produces, such as that of banging pipes. Much of this noise comes from the rapid condensation of the steam.

NATURAL-GAS PIPING SYSTEM

In many houses natural gas is used as a fuel in furnaces, hydronic boilers, and water heaters, as well as in ovens and range tops, gas lights, and barbecue grills. Natural gas is carried through large mains in the street. The natural-gas supply company runs individual lines from the main to the building, attaching a harness and gas meter. The plumber then installs the gas main inside the building and runs branches off it to the individual appliances.

Each natural-gas appliance must have an individual shutoff gas cock installed. This valve-type device allows the gas supply to each individual appliance to be shut off, permitting repairs or replacement without shutting off the main gas supply to the house.

Black steel pipe with brass and flexible gas connectors is usually used to supply natural gas (or propane gas) to specific appliances.

Remember: Never check a possible leak with a match or flame. Use a soapy solution and watch for bubbles.

Malleable black fittings, brass, and galvanized pipe and fittings may also be used. Natural-gas mains usually have a 1-inch inside diameter; branch lines usually have a ½-inch or ¾-inch inside diameter. Pipe dope is used on the pipe threads.

CUTTING AND JOINING PIPE

The method for cutting and joining pipe to other lengths of pipe and to pipe fittings depends on the type of pipe—cast iron, galvanized, copper, or plastic. However, with all types of pipe the first step is careful measurement.

There are several ways to measure pipe: from end to end, from end to center, from fitting to fitting, or in other ways, depending on the type of fitting being used. Measuring techniques are discussed in Chapter 6, Measurement and Conversion. Once you have measured carefully, mark the spot to be cut using a pencil, chalk, soapstone, or some other material that will show clearly on the pipe.

CAST-IRON SOIL PIPE

CUTTING CAST-IRON PIPE

Cutting cast-iron soil pipe on a job site is a skill requiring some practice and strength, depending on the tool used. Some pipes are fairly easy to cut, others more difficult.

Cast-iron pipe may be cut with a chain cutter, with a special abrasive cutoff machine, with a hacksaw, or with a chisel and hammer. The actual cutting is done by squeezing or pinching the cast iron with equal pressure on all sides so that it breaks apart. Correct placement of the cutting equipment is essential to a clean cut. If the cutting edge—whether a chain or wheel—is placed incorrectly, the pipe could be cut at an unacceptable angle, crushing it or making it impossible to join it properly to other pipes or fittings.

Using Chain Cutters. To cut cast-iron soil pipe with a chain cutter, lift the pipe and place the chain around it. Then lay the

(Figure 4-3) Cast iron pipe chain cutter

chain in the cradle. Tighten the chain until it is snug around the pipe. Then push the handle of the cutter (ratchet cutter), squeeze the handles together (squeeze-type cutter), or if the chain cutter is a hydraulic model with a foot pedal or hand bar, pump or push until the cast iron is cut. (See Figure 4–3.)

(Figure 4-4) Abrasive machine used to cut cast iron pipe

Using an Abrasive Machine. Place the marked pipe into the machine and pull down on the handle. The abrasive wheel cuts the pipe. When using this machine, follow all manufacturer's instructions and be sure to wear safety glasses. Abrasive material or small bits of iron can chip away and fly off. (See Figure 4–4.)

Using a Hacksaw. Use a coarse-toothed hacksaw to cut a shallow slot around the pipe on the line you have marked. Then use a hammer to tap along the groove so that the pipe breaks clean along the line. Be sure the saw blade is sharp. (See Figure 4–5.)

Using a Chisel and Hammer. Support the pipe on a good, solid bed, such as a block of wood. Position the chisel on the mark you have made and hit it with a strong hammer stroke. Turn the pipe and work the chisel around the line until the pipe breaks. (See Figure 4–6.)

JOINING CAST-IRON PIPE

There are several ways to join cast-iron pipe. Lead and oakum joints and gasket joints are two common ways, although gasket

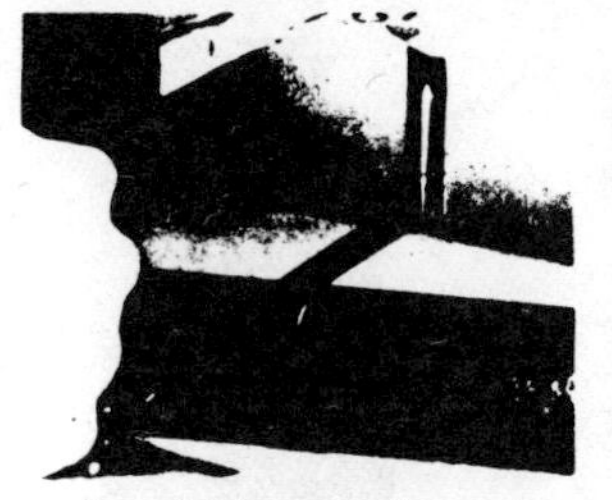

Use a hacksaw to cut groove all around pipe.

Tap along saw groove, and pipe will break along saw line.

(Figure 4-5) Using a hacksaw and hammer to cut cast iron pipe

joints may not be allowed by local plumbing code in certain situations (for example, for natural-gas lines). Special techniques using a neoprene sleeve and stainless-steel clamps are used to join no-hub cast-iron pipes.

Lead and Oakum Joints. One of the most common ways of joining cast-iron pipe is with lead and oakum joints. Follow these steps (refer to Figure 4–7):

Step 1. Insert the spigot end of one pipe into the hub end of another.

Step 2. Fill the space between the two pipe ends with oakum, which is a special type of rope. When inserting the oakum into the joint, be sure to twist it tightly.

(Figure 4-6) Using a chisel and hammer to cut cast iron pipe

Step 3. Yarn and pack the oakum into the joint using the proper tools. Usually a ball peen hammer is used to strike a special chisel-like tool known as a packing iron.

Step 4. Keep packing in the oakum up to about 1 inch from the top of the hub.

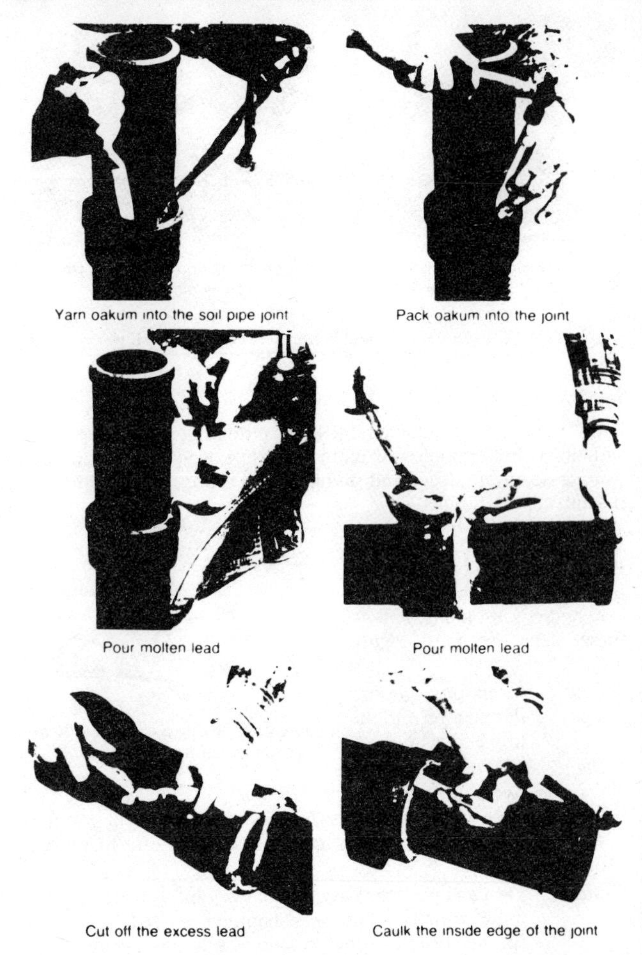

Yarn oakum into the soil pipe joint

Pack oakum into the joint

Pour molten lead

Pour molten lead

Cut off the excess lead

Caulk the inside edge of the joint

(Figure 4-7) Making a lead and oakum joint

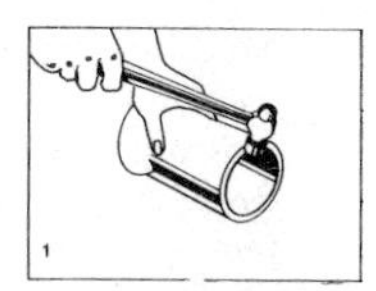

Peen the cut end of the soil pipe with a hammer to prepare it for insertion into the rubber compression gasket.

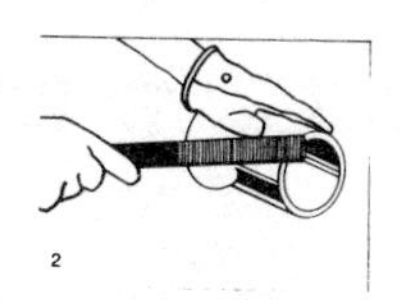

Rasp the cut end with a file to prepare it for insertion into the compression gasket.

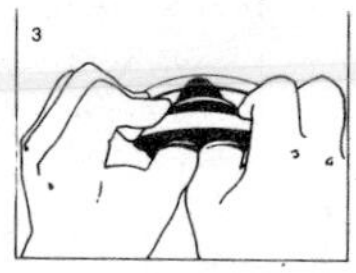

Fold the rubber compression gasket before inserting it into the soil pipe hub.

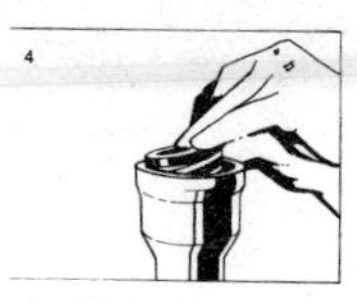

Insert the folded rubber gasket into the pipe hub.

(Figure 4-8) Inserting a rubber gasket in cast iron pipe

Step 5. Pour molten lead into the joint. Be sure there is no moisture in the joint. If moisture is present, it will turn to steam when the hot lead is poured in, and the lead in the joint may explode.

Step 6. Cut off any excess lead.

Step 7. Caulk the inside and outside edges of the lead using a special caulking iron.

Step 8. Trim or flatten edges for a professional look and to set lead for testing.

Gasket Joints. Rubber gaskets may also be used to join cast-iron soil pipe. But, remember that in some situations gasket joints are not allowed by the local plumbing code. (See Figure 4–8.)

To form a gasket joint:

Step 1. Smooth the outside edges of the cast iron.

Step 2. Lubricate the outside of the pipe hub or fitting and the inside area of the gasket.

Step 3. Fold the rubber compression gasket into the hub of the pipe.

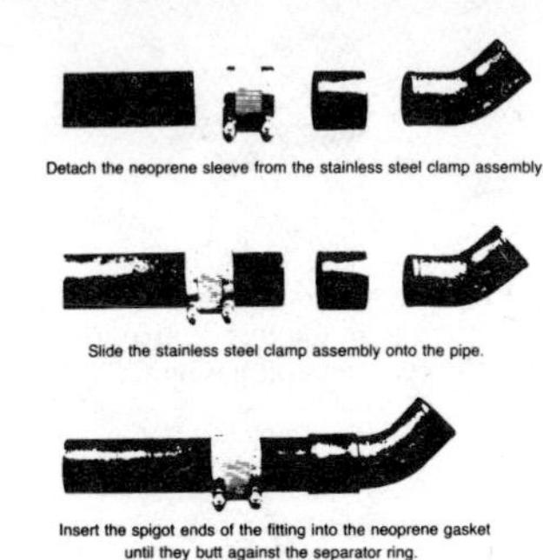

(Figure 4-9) Using a special fitting to join a no-hub pipe

Step 4. Make sure that the gasket has been seated properly.

Step 5. Insert the end of the pipe or fitting. Lubricate the spigot end of the pipe or fitting.

Step 6. Use a special assembly tool, a chain tool, or another tool to join the pipes or fittings tightly.

No-Hub Pipe Joints. No-hub cast-iron pipes can be joined using a special fitting. (See Figure 4–9.)

Step 1. Select the proper size fitting for the pipes to be joined.

Step 2. Make sure that the pipe ends are cut straight and that there are no sharp edges on the pipes or on the fittings.

Step 3. Separate the parts of the special gasket. It has a neoprene rubber sleeve, stainless-steel jacket, and screw clamps.

Step 4. Slide the stainless-steel assembly onto the pipe.

Step 5. Insert the end of the pipe into the neoprene gasket.

Step 6. Slide the stainless-steel jacket over the neoprene gasket.

Step 7. Use a torque wrench to tighten the steel clamps that hold the sleeve in place.

GALVANIZED AND BRASS PIPE

Galvanized steel, black pipe, and brass pipe are all available in the same pipe lengths and fitting sizes. Cutting, reaming, and threading these pipes is a skill that every plumber—and apprentice—must develop. Pipe sizes up to 2 inches are usually cut, reamed, and threaded on the job site; cutting and preparing larger pipe sizes usually requires special equipment and training and may often be done in the shop.

(Figure 4-10) A vise and special cutter can be used to cut brass or galvanized iron pipe

CUTTING

Although galvanized steel and brass pipe may be cut with a hacksaw, a special pipe cutter with a cutter wheel is usually used. (See Figure 4–10.)

Step 1. Place the pipe into a vise and tighten. Be sure you have measured and marked the location of the cut on the pipe.

Step 2. Select the pipe cutter and place the cutter wheel on the mark.

Step 3. Tighten the handle clockwise until the wheel presses against the mark on the pipe. Tighten lightly.

Step 4. Rotate the cutter back and forth. Gradually tighten the handle while turning the cutter 360 degrees around the pipe, but do not overtighten.

Step 5. Continue tightening the handle and rotating the cutter until the pipe is cut.

REAMING PIPE

(Figure 4-11) Reaming steel and brass pipe

Reaming is removing burrs (sharp edges) from the inside of pipe that has been cut. Burrs can snag waste material and lead to restricted flow and stoppages in a pipe. To ream steel and brass pipes, first select the reamer and insert it into the center of the pipe. Ratchet downwards (clockwise) until burrs are removed and the interior walls of the pipe are smooth. Then remove the reamer and clean out the steel or brass shavings. (See Figure 4–11.)

THREADING PIPE

After cutting and reaming a pipe, you must cut threads at the end of the pipe so that it may be screwed into steel and brass fittings. (See Figure 4–12.)

There are several different types of equipment used to make threads. Be sure to follow the manufacturer's operating instructions for each piece of equipment. In general, the following steps usually apply:

(Figure 4-12) Threading a galvanized iron pipe. Be sure to lubricate the die frequently

Step 1. Select the proper type and size of equipment for the size of pipe to be threaded.

Step 2. Lock the pipe firmly in a vise and place the threading equipment pipe guide on the pipe.

Step 3. Align the dies. Before beginning, be sure that the dies are sharp and not worn or broken.

Step 4. Ratchet downwards; this will press the die against the pipe, making the thread.

Step 5. Every couple of turns, lubricate the dies and threads. Plenty of oil is needed to protect the dies and threads and to ensure that the threads are clean and sharp.

Step 6. Make standard threads. Fewer threads will not allow pipe to enter the fitting as deep as it should for strength. Too many threads will allow the pipe to go deeper than it should into the fitting, and a leak could result.

Step 7. Clean the pipe threads and remove any steel or brass chips found inside the pipe.

TAPPING

Tapping is producing internal threads. It is done with a special tool, called a pipe tap, made of hardened steel. A pipe tap can be used not only to make internal threads but also to straighten damaged threads and to clean out rust and dirt inside a fitting that has been left in a wet or dirty area.

JOINING THREADED PIPES

Threaded pipes—galvanized, black, and brass—are usually joined to each other and to fittings by simply screwing them together. However, the joint should not be made dry. Be sure to use a pipe compound (pipe dope) on the male threads before screwing into the female threads. The compound lubricates the joint, prevents leaks, and makes it possible to unscrew the joint at a later date.

COPPER TUBING

Copper tubing is thin-walled and fairly easy to cut. Once cut, the tubing is joined to fittings and other lengths of tubing in several ways.

CUTTING COPPER TUBING

Copper tubing may be cut with a hacksaw, but it is better to cut it with special tubing cutters. These cutters come in several sizes, each designed for use on tubing of a specific diameter. For a complete discussion of copper-tubing cutters, see Chapter 1, Tools and Equipment.

JOINING COPPER TUBING AND FITTINGS

Copper tubing and fittings may be joined with compression joints or flare joints or by soldering.

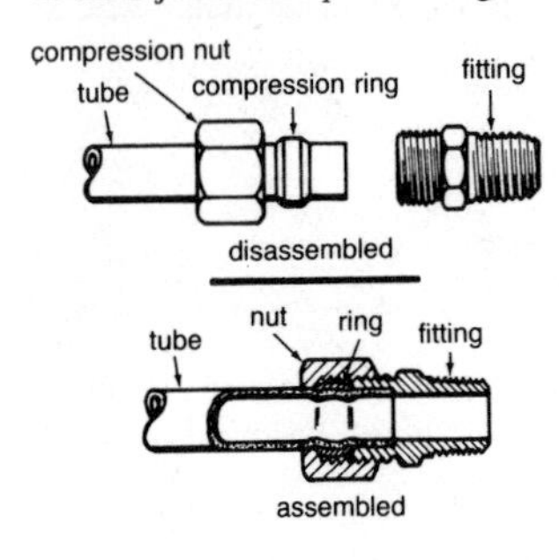

(Figure 4-13) Copper compression joint

Compression Joints. Compression joints make an airtight and watertight connection for all kinds of piping. A compression joint consists of a compression nut, a compression ring (ferrule), and some type of fitting, such as a male adapter. Two of the most common uses of compression joints are on plastic waste lines and on copper potable water-supply lines. (See Figure 4–13.)

To assemble a compression joint on a waste or water supply system:

Step 1. Measure the pipe (tubing) and mark the length needed.
Step 2. Cut the tubing at the mark.
Step 3. Make sure that there are no sharp edges on the inside or the outside of the tubing.
Step 4. Slide the compression nut onto the tubing.
Step 5. Slide the compression ring (ferrule) onto the tubing. Remember, once the ferrule is tightened, it cannot be removed.
Step 6. Insert the tube end into the fitting.

Step 7. Tighten the compression nut by hand until you can't turn it any more.

Step 8. Use an adjustable wrench or smooth-jawed pliers or wrench to tighten the connection. Do not overtighten. Overtightening can break or split the compression nut or strip the threads, causing a leak. (You can always tighten a connection a little if it leaks, but if it breaks or strips, you have to start all over again with new pipe and perhaps a new fitting.)

Some plumbers recommend using a little pipe dope on the threads of the connection to prevent friction from metal-to-metal contact.

Flare Joints. Flaring copper tubing is the strongest way to join lengths of copper tubing or to join tubing and fittings. Once the flare is completed, it provides an airtight and watertight connection. Copper tubing may be flared using a hammer-type flaring tool or a flaring block. The block is usually used for tubing sizes of ½-inch or less. The flaring tool is used for larger tubing.

To make a flare joint using a flaring tool:

Step 1. Cut the copper tubing to the desired length.

Step 2. Ream the interior of the tubing to prevent a ridge when you flare the tubing.

Step 3. Slide the flare nut over the copper tubing. This nut should move freely over the tubing.

Step 4. Insert the proper-sized flaring tool into the center of the copper tubing.

Step 5. Use a heavy hammer to drive the flaring tool until the flare is properly spread out. Be careful not to overflare the tubing. The flare nut must be able to slide over the flared ends onto the male threads of the fitting.

Step 6. Place the flared end against the fitting and tighten the flare nut by hand.

Step 7. Tighten the flare fittings. Be sure to use two wrenches when tightening.

To make a flare joint using a flaring block:

Step 1. Cut the copper tubing to the desired length.

Step 2. Ream the inside of the copper to prevent a ridge on the flared end.

Step 2. Ream the inside of the copper to prevent a ridge on the flared end.

Step 3. Slide the flaring nut over the copper tubing. The nut should move freely over the tubing.

Step 4. Place the copper tubing into the proper-sized hole. Move the copper tubing slightly above the flat surface of the flaring block and lock the flaring block.

Step 5. Tighten the compressor screw (clockwise) until it no longer turns.

Step 6. Remove the compressor screw and the flaring block.

Step 7. Slide the flaring nut over the flared end. The nut must be free to turn; if it is not, cut it off and start all over again.

Step 8. Use two wrenches to tighten the fitting.

Soldered Joint. Soldering is an old method of joining copper tubing and fittings, but it is still used by plumbers today. But before soldering any joint, be sure to check local plumbing codes and be aware of all federal regulations, which now prohibit the use on any potable water lines of any solder that contains lead. Solder that contains lead—60/40 solder and 50/50 solder—can, however, still be used on hot-water heating installations. (See Figure 4–14.)

To solder copper tubing:

Step 1. Measure and cut the copper tubing.

Step 2. Ream the interior walls of the copper tubing to be sure that they are free of all burrs.

Step 3. Clean the outside of the tubing.

Step 4. Clean the inside of the fitting to be used.

Step 5. Apply flux to the outside of the tube to the fitting depth.

Step 6. Apply flux to the inside of the fitting.

Step 7. Heat the copper tubing and the fitting. An acetylene or propane torch is usually used as the source of heat. The tubing and fitting should be heated to the working temperature of the solder being used—for example, 400°F to 450°F for 50/50 solder.

Step 8. Apply solder to the joint area. Be sure to cover the area completely and watch for the solder to flow into the fitting.

Step 9. Wipe the joint without moving the pipe or fitting.

Step 10. Wait until the joint cools, usually about one minute, before continuing with any work.

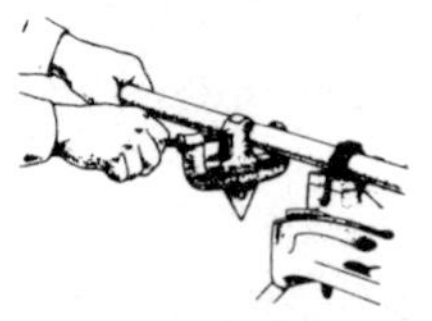

Cut the tubing with a roller cutter.

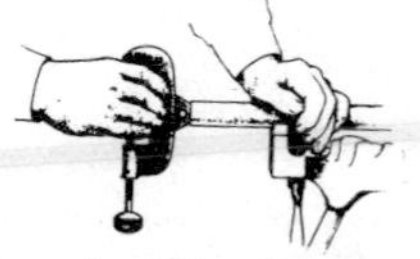

Ream the tubing. Do not over-ream or flare the tube.

Clean the tubing with sand cloth.

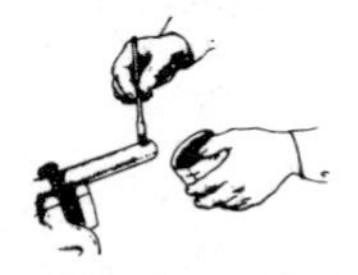

Flux the tubing. Apply just enough flux to cover the necessary surface.

Flux the fitting. If there are dirty spots on the cleaned surface, they will show under the flux.

Heat the tubing momentarily at first so that the heat transfers to the end of the tube. Point the flame slightly away from the fitting.

Heat the fitting. Be careful not to overheat it.

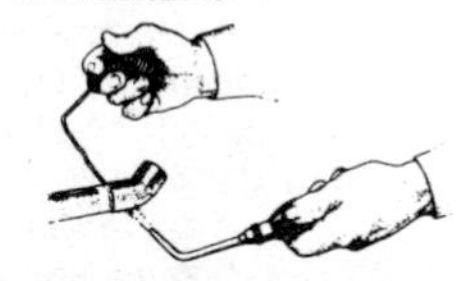

Apply solder to the heated joint. Heat the joint to the proper temperature and melt the solder by gently touching it to the juncture of the tube and the fitting.

Wipe the joint without moving the pipe or fitting. Let cool before continuing work.

(Figure 4-14) Soldering copper tubing

PLASTIC PIPE

Plastic pipe can be cut easily and joined to other lengths of piping or to fittings in several ways.

CUTTING AND REAMING PLASTIC PIPE

Most types of plastic piping can be cut using a saw, but special plastic-pipe cutters are also available. Deburring tools remove any raised edges on the inside of the piping. See Chapter 1, Tools and Equipment.

JOINING PLASTIC PIPE AND FITTINGS

Plastic water lines are usually joined using a flare technique or with insert fittings that are placed into the tubing. Solvent welding is also used.

To join plastic water lines using a flaring technique:

Step 1. Cut the plastic tubing to the correct length, using a special plastic-tubing cutter.
Step 2. Ream the inside of the piping to remove all rough edges.
Step 3. Slide the flare nut over the tubing.
Step 4. Insert the plug of the flaring tool into the tubing and rotate it 5 to 10 times.
Step 5. Place the flared end against the fitting and tighten the flare nut.
Step 6. Use two wrenches to tighten the connection.

To join plastic water lines using insert fittings:

Step 1. Cut the plastic tubing to the correct length.
Step 2. Ream the interior of the tubing to remove all burrs.
Step 3. Slip stainless-steel clamps on the two ends of the tubing.
Step 4. Push the insert fitting into the center of the plastic tubing until it reaches the proper depth. (See Figure 4–15.)
Step 5. Slide the stainless-steel clamps over the insert and tighten with a screwdriver, torque wrench, or nut driver.

To join plastic pipe using solvent welding:

Step 1. Select the proper cement for the job. Use ABS cement on ABS pipe and fittings, PVC cement on PVC pipe and fittings, and CPVC cement on CPVC pipe and fittings.

Step 2. Cut the plastic to the desired length. Be sure the cut is square. (See Figure 4–16.)

Step 3. Smooth the cut end of the pipe and ream the interior walls to remove all burrs.

Step 4. Clean the pipe and fittings with an all-purpose cleaner.

Step 5. Apply the cement to the pipe and fittings.

Step 6. Insert the pipe into the fitting and rotate ¼ turn.

Step 7. Wipe off excess cement.

Step 8. The curing time for plastic pipe varies with the outside temperature and the cement used. It is not good practice to fill or test plastic pipe and fittings joined by cement right away. It can take up to 24 hours for cement to cure. Check the cement manufacturer's recommendations.

(Figure 4-15) Joining plastic pipe with an insert fitting. Slide the stainless steel clamps over the ends of the two tube sections. Fit the plastic insert fitting into the tube. Once the clamps are in place around the insert, tighten them with a screwdriver.

INSTALLATION PRACTICES

Among the most common tasks of a plumber are installing appliances, faucets, fixtures, sump pumps, and related pieces of equipment. Although each appliance or fixture should be installed according to the specific instructions of the manufacturer, there are general procedures that apply in most situations. Every plumber and apprentice should be familiar with these.

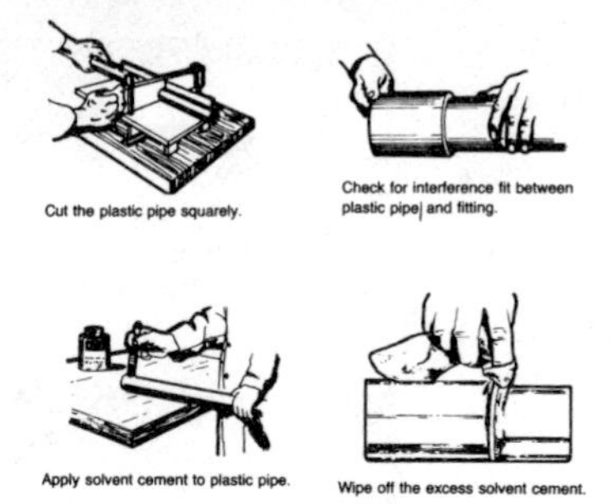

(Figure 4-16) Joining plastic pipe with a solvent weld

APPLIANCES

The appliances a plumber may be called upon to install are the dishwasher, hot-water heater, and washing machine.

DISHWASHER

Most dishwashers are installed under the kitchen counter either to the left or right of the kitchen sink. A dishwasher requires electricity, a hot-water line, and a waste line. Be sure the electrical hookup is completed before beginning to tie in the water and waste lines because once the plumbing installation is completed, the dishwasher cannot be moved to permit electrical wiring and hookup. After the electrical connection is made, follow the steps outlined below, adapting them as necessary to the specific manufacturer's instructions.

To install a dishwasher under a kitchen counter:

Step 1. Install a tee with a valve on the hot-water supply line under the kitchen sink. The valve allows the water to be shut off for future repairs or changes.

Step 2. Install a dishwasher tee between the duo (sink) strainer and the trap ("P" or "S") under the sink. (See Figure 4–17a.)

Step 3. Install drain hose, ⅝-inch I.D., running from the drain of the dishwasher to the trap. The drain hose must be installed higher than the top of the duo strainer to prevent the kitchen sink waste water from entering the hose and flooding the dishwasher itself. Do not kink the waste line while installing it. (See Figure 4–17b.)

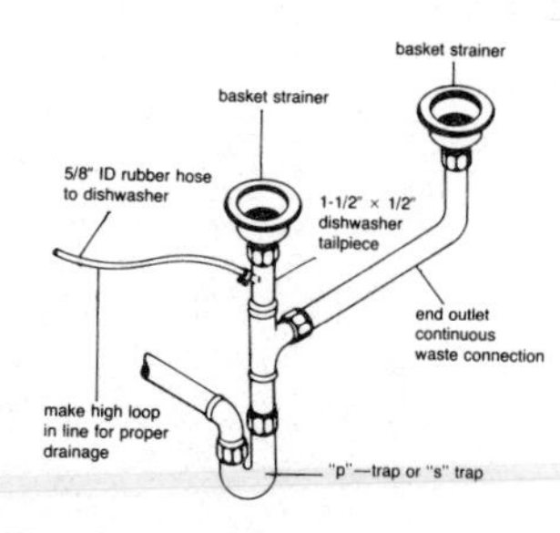

(Figure 4-17a) Strainer/trap section of a typical kitchen sink

Step 4. Install the hot-water line, ⅜-inch I.D., to the dishwasher electric water valve, which is located under the dishwasher. Do not kink the water line while installing it.

Step 5. To install the hot water line, use a ⅜-inch male pipe threaded connection made by flare fitting. Do not try to solder this connection since the heat may damage the electrical components in the valve.

A dishwasher can also be hooked up to a garbage disposal unit.

The installation of portable dishwashers (dishwashers not installed under the kitchen counter near the sink) depends on where and how the unit is to be used and may change with each particular situation. Follow the manufacturer's instructions.

GARBAGE DISPOSAL

A garbage disposal is installed under a kitchen sink in the area where the kitchen (duo) strainer is located. Screws are placed into a collar under the sink and tightened gently to make the section watertight. The base unit is twisted and locked into the collar under the sink. The top part of the disposer has a rubber or neoprene sleeve; when this is locked into the collar, there is an effective water seal. Plumber's putty may also be placed under the top section for sealing purposes. The waste connection is then hooked into a trap and the electrical connection made. A large amount of water should always be used to flush waste products through a garbage disposal system.

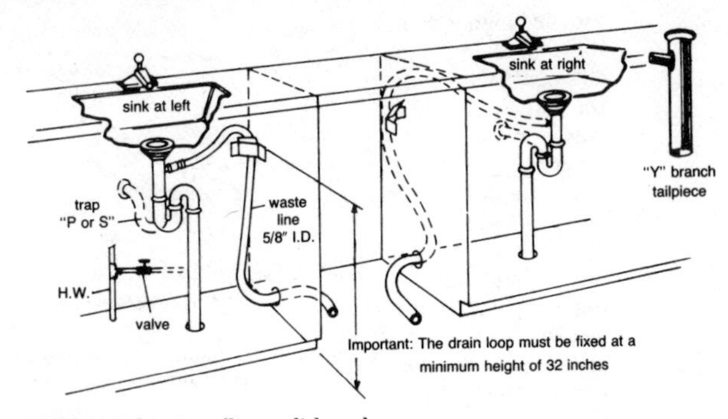

(Figure 4-17b) Installing a dishwasher

WATER HEATERS

Water heaters are installed to make the cold potable water—whether supplied to a building by the local water supply company or from a well—hot for use in kitchens, bathrooms, utility rooms, and basements. The cold water is heated either in an electric water heater, using 115 or 220 volts, or in a gas-fired heater fueled by natural gas, propane, or oil. It can also be heated in the furnace used for heating the building.

Electric Water Heaters. Electric water heaters come in many sizes, some as small as 6 gallons. They require no flue venting system and can be installed almost anywhere, even in a closet or under a sink. The unit must be wired according to state and local electrical codes. It will be supplied with cold potable water, which will be heated, and allowed to flow into pipes that will deliver it where needed throughout the house. (See Figure 3–2.)

Gas-Fired Water Heaters. Oil, propane, and natural gas water heaters require flue venting systems. Installing these units requires following all local fire and plumbing codes. However, basic installation procedures are usually simple. You need cold water to the unit, a fuel source to heat it, and a system of pipes to carry the heated water to the fixtures, faucets, and appliances where they are needed. (See Figure 3–4.)

To install a typical gas-fired water heater:

Step 1. Place the water heater in a convenient location.

Step 2. Supply the cold-water lines to the cold inlet of the heater unit.

Step 3. Place a valve on the cold-water inlet to allow ease of future repair or replacement. Usually a globe valve is sufficient for this installation.

Step 4. Place a valve on the hot-water side of the unit. This allows future repairs and replacements to be made more easily. It also allows the house water supply to be left on while the heating unit itself is being worked on.

Step 5. Install a temperature and pressure relief valve (required by plumbing codes in most states). This valve protects the heating unit from too-high temperatures or a buildup of pressure, both of which could damage the unit.

Step 6. Hook up the natural-gas, propane, or oil line, again following all code requirements.

Step 7. Light the pilot light according to manufacturer's directions. The usual steps involved are:

a) Turn the gas controller to the pilot position.
b) Depress the pilot button.
c) Light the pilot (next to the burner plate).
d) Release the pilot button.
e) Determine if the pilot flame is staying lit.
f) Turn the controller to "on."
g) Adjust the temperature to the desired setting.

Step 8. Install and connect flue pipes according to local codes.

Step 9. Adjust temperature settings. Avoid setting the temperature too high: it wastes fuel and money. (Also remember that excessively high temperatures can damage a heating unit.)

Step 10. Check that all manufacturer's instructions have been followed and that all state and local fire and plumbing code regulations have been satisfied.

WASHING MACHINE

Most washing machines are permanently installed in a utility room or basement. A washing machine requires an electric line, hot- and cold-water lines, and a waste pipe connected to a trap. A portable washing machine can usually be rolled to the kitchen sink and the waste water drained into the sink.

To install a washing machine:

Step 1. Hook up the large hose connection to the plumbing waste line.

Step 2. Hook up the two water-supply hoses to the hot- and cold-water distribution lines. *Do not cross lines.*

Step 3. Put the three-prong electrical plug into the wall receptacle.

Step 4. Make sure that there is proper grounding. Hook up the ground wire or other grounding device to protect the user from electric shock.

Step 5. Operate the machine according to the manufacturer's instructions.

SUMP PUMP

A sump pump is a type of rotary pump used to draw or lift water from a low level (sump) to a drain pipe. Sump pumps are sometimes used in basements that fill with water after a heavy rainfall. The pump allows the water to be removed from the floor and drained into the waste pipe, when the code permits it.

There are two types of sump pumps: centrifugal and submersible.

Centrifugal Sump Pump. In a centrifugal pump the motor is placed above the highest water levels. When the water rises in the pit, the float ball rises with it. When the float ball reaches a preset height, it turns on a switch located on the top of the motor. This turns on the pump suction and the water is drawn out of the pit and discharged through pipes to the outside. The lines should be run away from the basement so that the water does not recirculate into the nearby soil.

Centrifugal pump lines are installed with unions to allow for repair and replacement and also with check valves to prevent water from flowing back into the pit when the pump shuts off. Be sure that the electrical plug is grounded, and follow all manufacturer's instructions. (See Figure 4–18.)

Submersible Sump Pump. In a submersible pump, the motor can be under the water. Basically, a submersible pump works in the same was as a centrifugal pump and is installed in a similar way. Always be sure to use a three-prong electrical plug and to have the proper grounding. And, once again, follow the specific instructions of the manufacturer. (See Figure 4–19.)

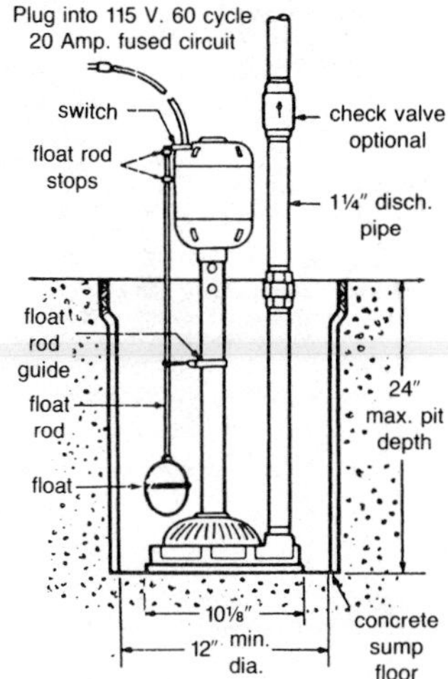

(Figure 4-18) Installing a centrifugal sump pump

FAUCETS

A faucet is really a special type of valve attached to the end of a pipe run to control the flow of water. There are many types of faucets, and each usually comes with a detailed parts list, diagram, and installation instructions. The plumber and plumber's apprentice should follow these specific instructions and save all manufacturer's information about replacement parts. There are, however, a few general guidelines that should be followed.

KITCHEN FAUCETS AND LAVATORY CENTERSETS

Kitchen faucets and lavatory centersets usually have two types of connections: a ½-inch iron pipe size (I.P.S.) connection and a ⅜-inch O.D. (outside diameter) copper tubing line connection.

To install ½-inch I.P.S. faucets:

Step 1. Insert the faucet through the holes in the sink.
Step 2. Thread the locknut finger tight onto the threaded pipe.

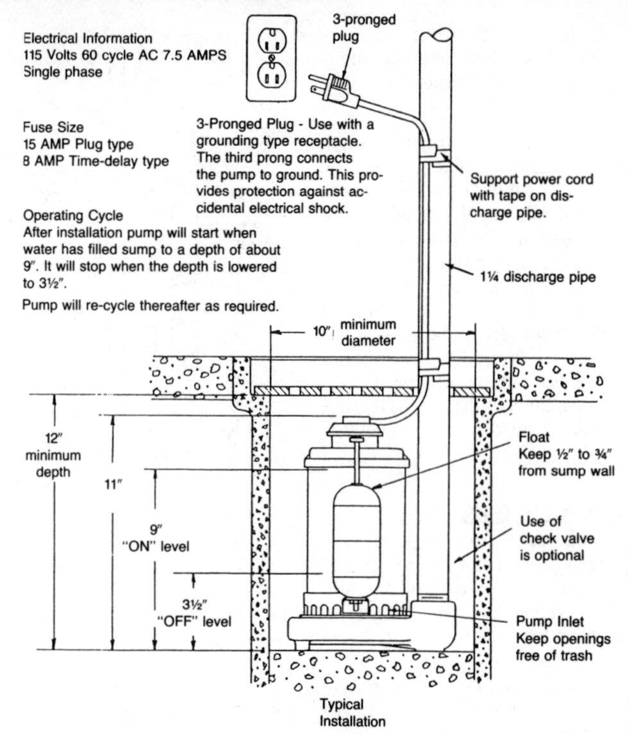

(Figure 4-19) Installing a submersible sump pump

Step 3. Tighten the connection with a basin wrench.

Step 4. Install the spray hose (if included).

Step 5. Install the flexible water supplies and tighten the slip-joint nuts for a watertight connection.

To install ⅜-inch O.D. tubing:

Step 1. Place the hot- and cold-water supply tubes into the center hole of the sink, if that is where the ⅜-inch O.D. tubes are.

Step 2. Lock the faucet in place by tightening the spacer.

Step 3. Carefully bend the ⅜-inch tubing to whatever angle is necessary to join it to the flexible water supplies coming from the chrome valve.

Step 4. Install ⅜-inch-to-⅜-inch compression unions with pipe dope and adjustable wrenches. Do not overtighten these fittings.

Figures 4–20 and 4–21 show the types of diagrams and installation instructions that are supplied with most faucets.

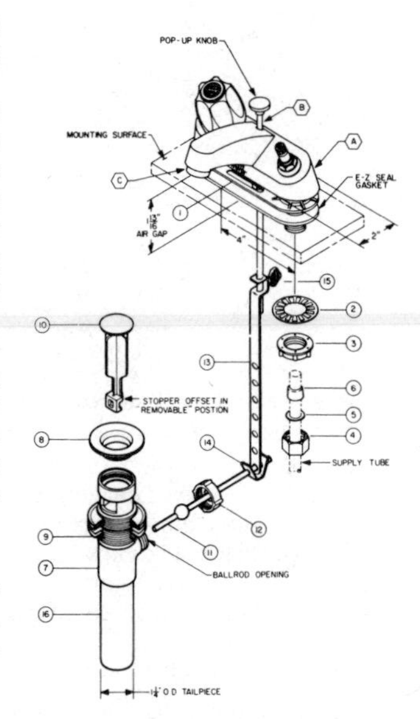

(Figure 4-20) Typical lavatory faucet. This and similar faucets come with parts described and detailed installation instructions.

FIXTURES

The installation practices for any fixture—kitchen sink, bathroom sink, bathtub, or other fixture—depend on the specific fixture and its placement in the building. Each figure comes with detailed parts and installation diagrams. The plumber should follow the general guidelines outlined below, adapting them to the specific directions provided by the manufacturer.

KITCHEN SINKS AND VANITY-TYPE SINKS

Kitchen sinks and vanity-type sinks set into countertops come either with sink rims or as self-rim types.

To install houdee-rim sinks in a countertop:

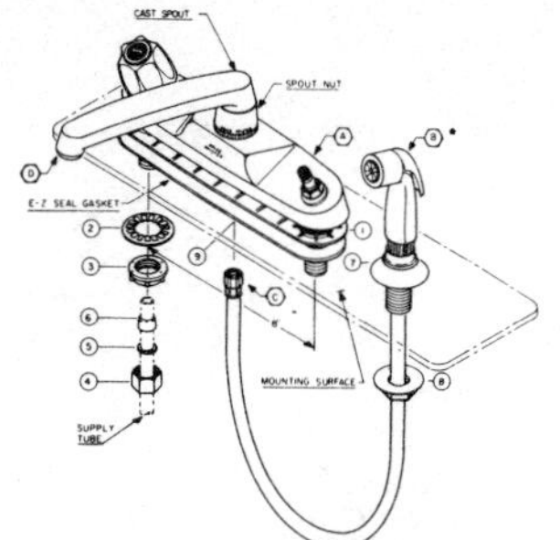

(Figure 4-21) Typical kitchen faucet with sprayer hose

Step 1. Place the houdee rim into the opening (if it is precut) on the top of the cabinet to see if it fits properly.

Step 2. Place sealing compound around the outside and inside of the houdee-rim frame. (This stops water from leaking into the cabinet.)

Step 3. Place the houdee rim around the sink being installed.

Step 4. Secure the sink from dropping into the cabinet and possibly damaging the sink. Have someone hold it or place a board across the top and wire it in place.

Step 5. Place sink clips under the sink and onto the houdee-rim edge. Tighten evenly. Do not overtighten sink clips because you could damage the porcelain finish of the sink.

Step 6. Clean excess sealing compound from the top of the sink.

To install one-piece (no houdee rim) sinks:

Step 1. Place sealing compound around the edge of the sink or basin.

Step 2. Position the sink in the countertop carefully.

Step 3. Make sure the basin is installed straight because once the sealing compound hardens, you may damage the sink if you try to turn it or move it in any way.

Step 4. Wait for the sealing compound to harden.

Step 5. Wipe or cut away any excess compound from all edges.

LAVATORIES

Lavatories or wall-hung basins come in various sizes and shapes. Most are made in colored enameled steel or vitreous china. These lavatories are mounted with special wall brackets.

To install a wall-hung lavatory:

Step 1. Remove the lavatory from the box or crate and find the specific wall bracket supplied with it.

Step 2. Place the wood backing behind the sheetrock or tile. This backing is necessary for screwing in the steel or cast-iron brackets. The use of expansion bolts or toggle bolts is not recommended since they will eventually pull out of the wall, leading to considerable damage.

Step 3. Check the measurements. Be sure that once the lavatory is hung, the flood level of the fixture will be the standard height off the floor—usually 31 inches.

Step 4. Bolt or screw the brackets into the backing and install the lavatory.

BATHTUBS

Bathtubs must be installed and tested during the roughing-in stage—to be sure that proper connection of water piping and waste piping is possible. All bathtubs, regardless of their size, shape, or material, require the same general installation practices but, as always, be sure to follow the directions of the manufacturer.

To install a bathtub:

Step 1. Do all preinstallation work, using roughing-in dimensions given on the architect's or engineer's plans or the mechanical plumbing diagrams. This includes making clearance holes for drainage pipes, space for traps and valves, and basic framing and space for faucets and shower heads. Be sure that the height and lateral position of each pipe is cor-

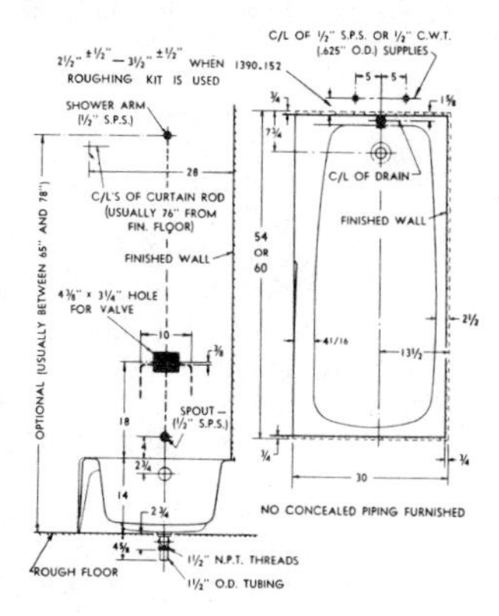

(Figure 4-22) Roughing-in dimensions for a typical bathtub

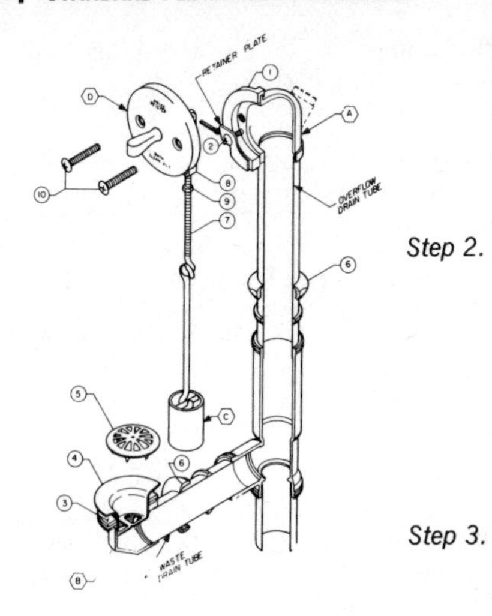

(Figure 4-23) Typical trip lever waste overflow

rect. Do any framing or bracing that is necessary to ensure proper positioning and leveling of the tub. (See Figure 4–22.)

Step 2. Check to see that the tub to be installed is in perfect condition, with no chips or cracks. (This check *must* be done; it is extremely difficult and costly to remove and replace a tub.)

Step 3. Install legs under the rim of the bathtub. Usually three legs and a long (5-foot) two-by-four placed over them will prevent the tub from rocking when it is filled and in use. (If a bathtub rocks, it will loosen the grout around the rim of the tub and tile, causing water leakage in the wall.)

Step 4. Install the tub. Be sure that it is level.

Step 5. Install the waste and overflow, and hook up to the waste trap. Follow all the manufacturer's instructions. (See Figure 4–23.)

Step 6. Using water, check for leaks. (However, many bathtub leaks only appear later, often as spots on the ceiling below.)

Step 7. Cover the bathtub with newspaper or a plastic liner to minimize the chance of chipping or other damage while plumbers and other construction workers complete work in the bathroom. (A tub is very expensive to remove and replace, so all precautions to minimize the chance of damage should be taken.)

SHOWERS

Showers come in all different shapes and sizes; some are one-piece units, others have a shower base only. Showers should be placed during the roughing-in phase so that all the waste piping and water piping is properly connected. Once the shower is set in place, waste lines should be hooked to a trap and then the system tested. Detecting any possible leaks—and thus poor connections—now will prevent later damage to walls and ceilings.

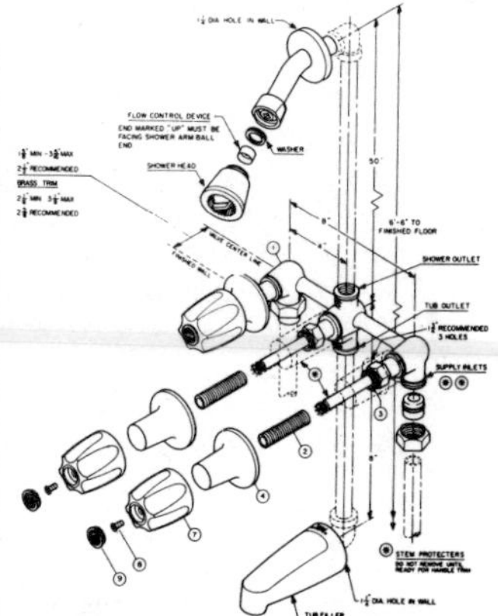

(Figure 4-24) Typical shower diverter

Years ago shower stalls were often set in cement or the bases made of lead, and then cement and tile were placed over them. These shower units were difficult to repair. Today most shower units are made of fiberglass. Many come as complete packages and are very easy to install. Just follow the manufacturer's instructions. (See Figures 4–24 and 4–25.)

TOILETS (WATER CLOSETS)

Toilet bowls come in various shapes and sizes. Two standard designs are the round and the elongated front. The round front is good for close spaces; the elongated requires more space but is required in most public restrooms for sanitary reasons.

To install a toilet bowl:

Step 1. Install floor flanges, if not already installed.

Step 2. Screw wood screws into the wood flooring underneath the floor flange.

Step 3. Place water closet bolts into the holes on each side of the floor flange, usually 12 inches from the wall.

Step 4. Place wax seal onto the bottom of the toilet bowl around the large hole.

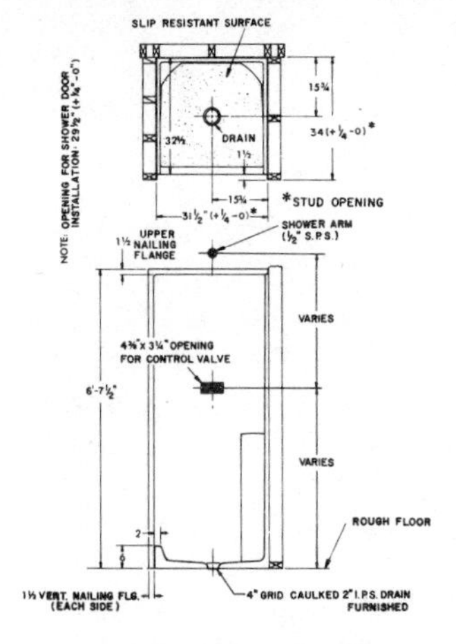

(Figure 4-25) Roughing-in dimensions for a typical one-piece fiberglass shower enclosure

Step 5. Pick up the toilet bowl and hold it directly over the floor flange and align the toilet bowl holes over the bolts. Lower the bowl gently until the wax seal and the floor flange come together.

Step 6. Rock the toilet bowl gently from side to side and from back to front. When the bowl comes in contact with the floor or tile, stop.

Step 7. Put the nuts and washers onto the bowl bolts and tighten. Do this from one side, then from the other side. Do not tighten one side only. You could snap or break the toilet bowl edge if you do.

Step 8. Use a hacksaw or other tool to break the top of the bolts off. Place caps on top of the bolts for a professional finish.

Step 9. Install the toilet tank according to the manufacturer's instructions, which are usually found in the packet of nuts and bolts inside the tank.

POTABLE WATER SERVICE

Potable cold water is supplied to a building either from a water-supply company's main in the street or from a well somewhere on the property. Every plumber and apprentice should be familiar with some basic facts concerning these systems and with the underground water lines used in either system.

WATER-MAIN SERVICE

Water-supply companies may be run by the government (municipal or town) or be privately owned. They bring water through a large pipe, called a main, to the street. Then, usually for a fee, the company will bring the water line—a water-service line—to the property line. The company will usually leave a curb box and valve that allows the water service to be turned on or off. A meter is also installed so that the amount of water consumed in the building can be measured. The measured amounts are then used to calculate the building owner's water bill.

Although local and state codes do vary, most water-main and water-service lines are installed underground below the frost line. The water piping is usually galvanized or brass pipe, copper tubing (L or K type), or plastic (PE) tubing. (See Figure 4–26.)

The installation of water lines is strictly governed by code regulations and state and local ordinances. Permits are required to install these lines. As always, be sure to check all applicable regulations.

WATER FROM A WELL

Before installing any well system, be sure to contact a reputable well driller who is familiar with the property and with all local

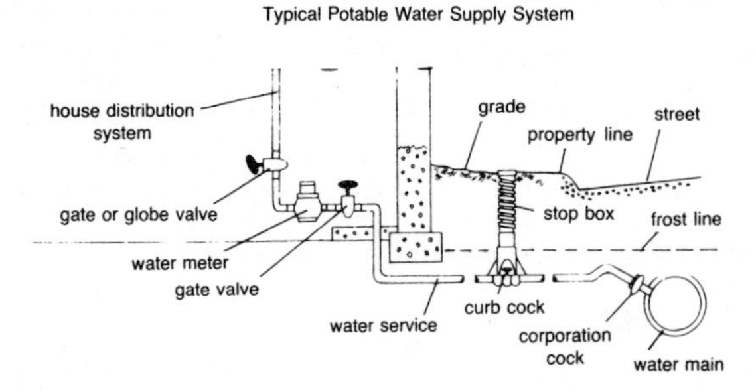

(Figure 4-26) Typical potable water supply system

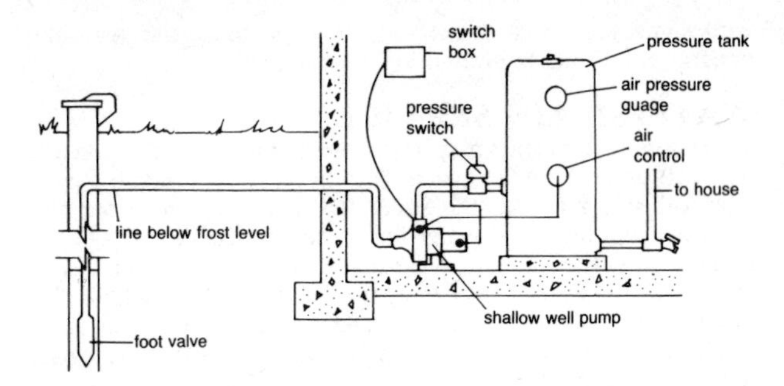

(Figure 4-27) Typical well

laws governing wells. Many states, counties, and townships have laws requiring that wells be located a considerable distance (sometimes as much as 100 feet) from any septic system to prevent any possible contamination of potable water sources. A reputable well driller will advise you on these matters and other important factors in the placement and design of a well. Remember to obtain advice and become familiar with the local laws and with the property before beginning to drill or construct any buildings. (See Figure 4-27.)

TYPES OF WELLS

Shallow Wells. Shallow wells—wells down to 30 feet deep—can be dug by hand or drilled. They are easy to install but have one major disadvantage: they are easily contaminated by surface water or septic systems.

Deep Wells. Deep wells, more than 30 feet deep and often drilled to considerable depths, are more difficult to prepare. They are, however, not easily contaminated.

TYPES OF PUMPS

There are several types of pumps that are used with wells. The choice of which to use depends on the depth of the well, the terrain of the property, and other factors. Installation instructions are supplied with the pumps.

Jet Pumps. Jet pumps are used to push and draw water to the surface. They are installed in a two-pipe system within the well.

Submersible Pumps. Submersible pumps are placed inside the well below the waterline. They are sealed so that water cannot damage the motor.

Reciprocating Pumps. Reciprocating pumps are simple pumps usually used only in shallow wells.

LINES FROM THE WELL

Water lines leading from the well should be installed below the frost line to prevent freezing. Galvanized or brass pipe, copper tubing, or plastic (PE) piping are usually used.

FITTINGS FOR ALL UNDERGROUND WATER LINES

As a general rule, you should avoid placing any fittings underground, especially in regions where the ground can freeze during the cold months. Fittings under concrete slabs should also be avoided. Fittings are the weakest point in any water line. Trying to repair or replace broken or leaking fittings underground or under concrete is extremely difficult and expensive.

The fitting at the curb box on a water line is usually an iron threaded joint (male or female) or a flare joint. Couplings are sometimes used to join two pipes or tubing materials to cover a long distance. A flare joint is strongest and should be used over a soldered joint if at all possible. If PE tubing is being used, make every effort to install one continuous line of tubing underground. Stainless-steel clamps and inserts (ways of joining plastic tubing) have been known to blow apart under severe pressure from ground pressure caused by the ground settling.

Water lines usually do not need pitch, and it is usually not required by state and local codes (as it is for waste piping).

REPAIR TECHNIQUES

Plumbers probably spend most of their time doing repair work—fixing dripping faucets, clearing waste stoppages, repairing leaking pipes and toilets. Although the specific practices to be followed depend on the specific fixture and the manufacturer's instructions, there are some basic guidelines that all plumbers should be familiar with. Also check Chapter 3, Preventive Maintenance and Troubleshooting, for hints on diagnosing problems and ways of solving them.

FAUCET LEAKS

Most faucets come with detailed parts diagrams, installation instructions, and repair advice. This information should be saved so that you can order parts for replacement.

First, determine the source of the leak. Is it coming from the aerator, swivel joint, stem area, or other part? Decide whether the repair requires that the water to the faucet be shut off. If it does, shut off the valve supplying the fixture.

If the drip is a simple problem with the aerator, the water need not be turned off. Simply clean or replace the aerator. (This type of leak will usually stop, but it can be annoying and can ruin a fixture by causing rust stains.)

Check all washers and brass seats (if present), and replace them if they are worn or defective. If the faucet is a cartridge type and the cartridge is found to be defective, replace it.

If the leak is coming from the stem area of the faucet, tighten the packing nut or replace the O-rings on the stem. Rarely does the stem need replacing, but replacement can sometimes be the only way to resolve a particular problem.

Kitchen sink faucet spouts swivel from one side to the other. If water leaks from the swivel joint, remove the spout and replace the O-ring or packing.

PIPE LEAKS

The repair of pipe leaks depends on the piping material and the method used to join pipe lengths and fittings.

CAST-IRON PIPE LEAKS

Cast-iron pipe joined using lead and oakum usually does not leak after the initial installation and testing. If a leak does occur, it often indicates that the hub and joint are under high water pressure, such as that resulting from a waste stoppage. If this is the case, the stoppage must be cleared.

If a lead and oakum joint leaks and the cause is not a waste stoppage, you can repair the leak by gently tapping the lead ring. You can also apply epoxy to the hub area.

BRASS AND GALVANIZED PIPE AND FITTING LEAKS

Leaks in a galvanized or brass piping and fitting system usually occur when the system is first tested. Any threaded joint that leaks should be taken apart. Check the threads, look for sand holes in the fitting, apply pipe dope or a sealing compound, and retighten the pipe and fittings, using two wrenches. Older pipes might need to be replaced, rather than repaired.

COPPER TUBING AND FITTING LEAKS

If the copper tubing and fittings have been joined using a soldered joint, cut out the leaking portion, replace it with a new fitting, and resolder. Use the old fitting only if you are sure it is clean and still usable. Copper joints made using flaring techniques seldom leak. Check the flange nut first. If a leak persists, cut off the old flare, use a new fitting, and remake the joint.

PLASTIC PIPE AND FITTING LEAKS

Leaks in plastic pipe and fittings should he handled the same as leaks in a copper system. Undo the joint, remove the fitting, replace it, and redo the joint, using a flaring technique or special cement. Temporary repairs (or easy fixes) usually create problems later.

TOILET LEAKS

The toilet is a plumbing fixture that often needs repair. Leaks may occur in several spots in and under the toilet bowl and toilet tank.

If a leak occurs in the water valve underneath the toilet tank, tighten the packing nut on the shut-off valve.

If water leaks out from the ferrule and compression nut-top of the valve located under the toilet tank where the flexible toilet supply tube is inserted, gently tighten the ferrule and compression nut. If this does not stop the leak, you may need to replace the flexible water-supply tube.

If water drips from the water tube into the bottom of the toilet tank, gently tighten the large compression nut under the tank. If this does not stop the leak, you may have to remove the nut and check the neoprene or rubber gasket washer that makes the watertight seal. If necessary, replace the washer.

If water leaks from under the toilet bowl when you flush, gently tighten the two toilet bolts on either side of the bowl. This may reset the wax seal. If not, pick up the toilet bowl and remove the old wax seal. Install a new wax seal. Do not try to reuse the old one.

If the toilet leaks under the tank and on top of the bowl when you flush, gently tighten the two bolts under the tank between the tank and the bowl. If this does not stop the leak, remove the tank from the bowl and replace the closet spud gasket.

Leaks inside the toilet tank itself usually indicate that the float ball, ballcock assembly, or other part needs to be replaced.

VALVE LEAKS

A leak coming from a valve usually indicates that a packing nut has loosened with use. Gently tighten the packing nut until the leak stops. Do not overtighten. If the joint area near the valve is leaking, you probably should undo the joint and replace it, using the same technique that was used during the initial installation.

WASTE-PIPE STOPPAGES

Stoppages in waste pipes and traps are common.

Kitchen Sink Stoppage. To correct a kitchen sink stoppage, use a plumber's helper or small snake to dislodge the stoppage. Remove the trap under the sink (be sure to place a pail under the trap area) and clean it. Most stoppages are found in the trap.

Lavatory, Vanity, and Bathtub Stoppages. Remove the pop-up plug and clean. Remove the trap (if possible) and clean using a small wire snake to eliminate the stoppage. Most often hair and soap particles are responsible for the stoppage.

Shower Drain Stoppages. Remove the top plate from the drain and clean the inside area of the pipe. Often soap and hair reduce the inside diameter of the pipe, and water backs up into the shower. Use a small flexible drain snake to go down into the drain pipe.

Toilet Stoppages. Try a plumber's helper's up-and-down motion to dislodge the stoppage. But if the stoppage is in the bowl, use an auger to snare and remove the object. Most objects will not go through the bowl and must be removed. Trying to push them through (if it is even possible) may only lead to a stoppage farther down the waste line.

Stoppages in the Main Waste Line. These stoppages can result from large objects in the line, sewer or septic system problems, or tree roots growing into the pipes. Usually these stoppages are handled by professional waste-removal (cesspool or septic system) companies.

ROUGHING IN

Roughing in is a stage of construction in either a new building or an extension of a building. It is the earliest stage of plumbing installation and involves measuring, cutting, joining, and installing basic piping systems.

Roughing in is sometimes divided into two stages. The first brings water and sewer lines inside the building foundation. The second involves the installation of all piping that will be enclosed in the walls of the finished building.

Although an apprentice plumber will not be involved in the basic roughing in and design of an entire plumbing system for a building, he or she should have some understanding of what is involved.

ROUGHING IN BASIC SYSTEMS

Roughing in basic systems requires the ability to read blueprints and use a ruler for scaling and a knowledge of basic piping materials and fittings. Plumbers must know which materials and fittings should be used in particular situations. There are several steps and procedures that should generally be followed:

- Be sure there is an accurate drawing of the plumbing system to be installed and that the drawing is submitted with the permit application to the local authorities. A fee is usually required for inspections.
- Determine how waste material is to be handled. Is there a sewer in the street so that waste piping can be hooked to it? Or will waste be handled by a cesspool or septic tank system?
- Find out if there is a water main in the street or if a well is to be used to supply potable water.
- Install underground water lines from the water main, water-service line, or well line to the building. Install the waste lines from the building to the sewer or cesspool/septic system. Be sure that all government rules and plumbing codes are obeyed. Underground water lines, for example, usually have to be installed below the frost line, and waste lines must be installed at a certain pitch (usually ¼ inch per foot) to ensure proper flow.
- Check the blueprints and layout for all waste, water, and vent pipe locations.
- Install all water, venting, and waste piping and fittings using the proper materials. (Check local codes.)
- Join all pipes and fittings correctly.
- Test all pipes and fittings for leaks and defects. Replace any defective connections and correct any other problems.
- Be sure that all pipes and fittings are located in the correct spot for hookup to the various fixtures and appliances.
- Install bathtubs and showers before sheet rock and tile is installed.

- Be sure that all water and waste piping is protected during construction to avoid nails being driven into water or waste pipes or having foreign particles enter the pipes.
- Prepare for the roughing in and installation of fixtures and appliances.

ROUGHING IN FIXTURES

The location of waste and water lines varies from fixture to fixture. Not only must the connection points on the fixtures and the basic pipes in the walls be aligned, but also the location of all studs, beams, and other building parts must be considered.

Each fixture differs in measurements. There are many types of sinks, toilets, tubs, showers, and appliances. The general roughing dimensions used when installing basic plumbing lines must be adjusted to accommodate the specific fixture. All of this takes careful measuring and layout work by an experienced plumber.

We will briefly mention some general roughing-in dimensions for waste and water piping for several fixtures. But, as always, follow all manufacturer's diagrams, measurements, and instructions.

Water Closet (Toilet). The water closet rough measurement for waste pipe is 12 inches from the wall (add tile and/or sheet rock measurements) to the center of the waste connection of the 3- or 4-inch floor flange. Toilet bowl water-supply lines (only cold water) are roughed usually on the left-hand side of the toilet at about 8¼ inches. (See Figure 4–28.)

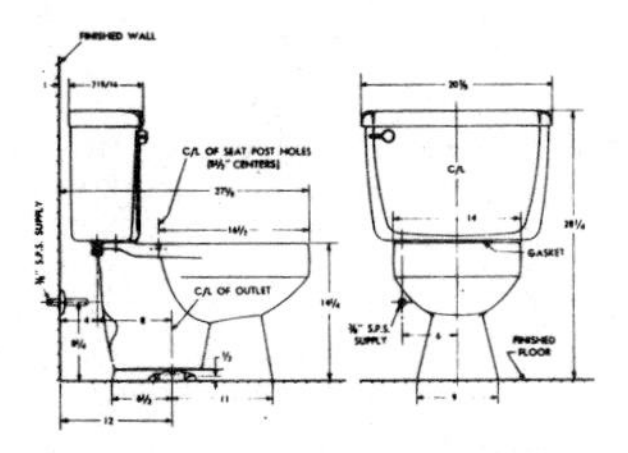

(Figure 4-28) Roughing-in dimensions for a typical toilet

Basin or Lavatory. Basin or lavatory waste connections are usually about 16¾ inches from the finished floor to the center of the 1¼-inch or 1½-inch waste pipe. The height does, however, vary with the particular basin or lavatory selected. Water-supply pipes are usually ½ inch and are roughed approximately 21 inches off the finished floor. (See Figure 4–29.)

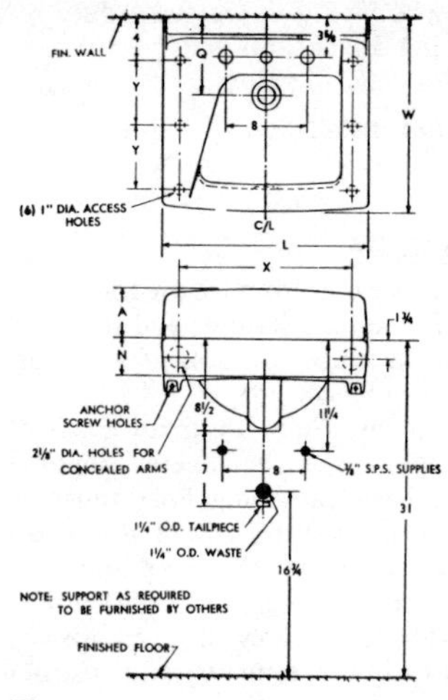

(Figure 4-29) Roughing-in dimensions for a typical wall-hung lavatory

Kitchen Sink. Kitchen sink waste connections vary, depending on whether or not a garbage disposal is to be installed and other factors. Normal installation is 17 inches from the finished floor, but again it varies with the equipment chosen. Water pipe size is most often ½ inch roughed in at 21 inches off the finished floor.

Bathtub. To rough in the waste and overflow, bathtub trap and waste pipe, the plumber usually needs to cut out a floor area at the head of the bathtub. This area is cut out by using the front center of the tub as a guide—the cut area should be 4 inches on both sides of the center mark and 10 inches deep. This cutout will allow the bathtub waste trap to be installed and leave room for it to be tightened in case of future leaks. (See Figure 4–22 and Figure 4–23.)

PART THREE

PLUMBING FUNDAMENTALS

5

Basic Mathematics and Formulas

Mathematics is used every day by the plumber and plumber's apprentice. All plumbers must be familiar with the basic terminology and symbols of mathematics, with basic arithmetic processes, and with concepts and formulas in geometry and trigonometry in order to pursue daily tasks. In addition, plumbers should be familiar with special formulas available to aid them in calculating pipe offsets, sloping, and other problems specific to plumbing situations.

This chapter lists some basic terms in mathematics, summarizes facts and formulas in arithmetic, geometry, and trigonometry, and then discusses formulas—with examples on how to use them—of particular interest to plumbers. Tables are given where applicable; however, check the appendixes for additional tables and useful data.

DEFINITIONS

abstract number a number that does not refer to any particular object.

acute triangle a triangle in which each of the three angles is less than 90 degrees.

altitude of a triangle a line drawn perpendicular to the base from the angle opposite.

angle the difference in direction of two lines proceeding from the same point called the *vertex*.

area the surface included within the lines which bound a figure.

arithmetic the science of numbers and the art of computation.

base of a triangle the side on which a triangle is supposed to stand.

board measure a unit for measuring lumber, the volume of a board 12 inches wide, 1 foot long, and 1 inch thick.

circle a plane figure bounded by a curved line, called the *circumference*, every point of which is equally distant from a point within, called the *center*.

complex fraction a fraction whose numerator or denominator is itself a fraction.

cone a body having a circular base and a convex surface that tapers uniformly to the vertex.

cubic measure a measure of volume involving three dimensions—length, width, and thickness (depth).

cylinder a body bounded by a uniformly curved surface, its ends being equal and parallel circles.

decimal scale a scale in which the order of progression is uniformly ten.

diameter of a circle a line that passes through the center of a circle and is terminated at both ends by the circumference.

diameter of a sphere a straight line that passes through the center of the sphere and is terminated at both ends by its surface.

equilateral triangle a triangle that has all its sides equal.

even number a number that can be exactly divided by two.

exact divisor of a number a whole number that will divide a number without leaving a remainder.

factors two or more quantities which, when multiplied, produce a given quantity.

factors of a number numbers which, when multiplied, make that specific number.

fraction a number which expresses part of a whole thing or quantity.

geometry the branch of mathematics that treats space and its relations.

greatest common divisor the greatest number that will exactly divide two or more numbers.

hypotenuse of a right triangle the side opposite the right angle.

improper fraction a fraction in which the numerator equals or exceeds the denominator.

isosceles triangle a triangle which has two of its sides equal.

least common multiple the lowest number that is exactly divisible by two or more numbers.

measure the extent, quantity, capacity, volume, or dimensions ascertained by some fixed standard.

mensuration the process of measuring.

number a unit or collection of units.

odd number a number which cannot be exactly divided by two.

parallelogram a quadrilateral which has opposite sides that are parallel.

percentage the rate per hundred.

perimeter the distance around a figure, the sum of the sides of a figure.

perpendicular of a right triangle the side that forms a right angle to the base.

proper fraction a fraction in which the numerator is less than the denominator.

pyramid a body having for its base a polygon and, for its other sides or facets, three or more triangles that terminate in a common point called the vertex.

quantity an aspect that can be increased, diminished, or measured.

radius of a circle a line extending from the center of a circle to any point on the surface.

rectangle a parallelogram in which all angles are right angles.

right triangle a triangle which has a right angle (90 degree).

scale the order of progression on which any system of notation is founded.

scalene triangle a triangle on which all sides are unequal.

sphere a solid body bounded by a uniformly curved surface, all the points of which are equally distant from a point within called the center.

square a rectangle whose sides are equal.

trapezoid a quadrilateral that has two parallel and two oblique sides.

triangle a plane figure bounded by three sides and having three angles.

uniform scale a scale in which the order of progression is the same throughout the entire succession of units.

unit a single thing or a definite quantity.

varying scale a scale in which the order of progression is not the same throughout the entire succession of units.

BASIC ARITHMETIC

It is not in the scope of this book to explain the basic processes of arithmetic: addition, subtraction, multiplication, and division. However, we should like to point out that the plumber and apprentice

TABLE 5–1

BASIC MATHEMATICAL SYMBOLS	
= equal to	× times, multiple by
+ plus, add to	÷ divide by
− minus, subtract from	° degree
π pi, equal to $\frac{22}{7}$ or 3.1416; used in calculating the circumference and area of a circle	
( ) parentheses; all numbers and the result of all operations within parentheses are considered one quantity. If a series of operations are to be performed, those enclosed in a parentheses must be done first and the result used as a single quantity in all other operations. For example, in the problem $75 + (3 \times 41) - 20$, first multiply 3 by 41. Then using the result—123—go on to the rest of the problem: $75 + 123 - 20 = 178$. (You *cannot*, for example, proceed as follows: $75 + 3 = 78$, $78 \times 41 = 3198$, $3198 - 20 = 3178$.)	

must be able to perform all of these basic operations accurately and quickly. It is absolutely essential in calculating pipe and fitting sizes and positions and in every other aspect of daily plumbing work. A knowledge of basic arithmetic is also necessary in ordering supplies and calculating costs and charges. (See Table 5–1.)

A FEW REMINDERS ABOUT BASIC PROCESSES

Remember that only like units can be added or subtracted. Just as you can't add apples and bananas, you can't add feet and inches, unless you change all the quantities to the same base unit. For example, if you have a series of measurements such as 2 feet, 8 inches, 11 inches, and 4 inches, and need to add them to find the total length, you must first change the 2 feet to 24 inches (there are 12 inches in 1 foot) and then add. The result—in this case, 47 inches—can then be left as is or changed into another unit—in this case, 3 feet 11 inches.

In multiplication and division, there are a few additional points to remember. When two linear units are multiplied—for example, feet by feet, or inches by inches—the product is in square units—square feet or square inches. If you wish to know the area of a rectangle that measures 3 feet by 6 feet, you multiply 3 by 6 and the result is 18 square feet. Conversely, if you know that the area of a rectangle is 36 square feet and that one side is 3 feet, you can divide to find that the other side is 12 feet (not square feet).

FRACTIONS

A fraction indicates a part of a whole. To add or subtract fractions, remember that their denominators (the lower part of the fraction) must be the same. Add or subtract their numerators.

$$\begin{array}{r} \frac{6}{10} \\ +\ \frac{3}{10} \\ \hline \frac{9}{10} \end{array} \qquad \begin{array}{r} \frac{7}{10} \\ -\ \frac{4}{10} \\ \hline \frac{3}{10} \end{array}$$

If the denominators are not the same, you must change the fractions (without changing their values) so that the denominators are the same.

$$\begin{array}{rcl} \frac{1}{2} & = & \frac{2}{4} \\ +\ \frac{1}{4} & = & \frac{1}{4} \\ \hline & & \frac{3}{4} \end{array} \qquad \begin{array}{rcl} \frac{1}{5} & = & \frac{2}{10} \\ -\ \frac{1}{10} & = & \frac{1}{10} \\ \hline & & \frac{1}{10} \end{array}$$

To multiply fractions, simply multiply the numerators and the denominators.

$$\frac{1}{5} \times \frac{3}{4} = \frac{3}{20}$$

To divide fractions, invert the divisor and then multiply.

$$\frac{4}{5} \div \frac{1}{2}$$

$$\frac{4}{5} \times \frac{2}{1}$$

$$\frac{4}{5} \times \frac{2}{1} = \frac{8}{5}$$

An improper fraction, such as the division answer $\frac{8}{5}$, should be converted to a mixed number—in this case $1\frac{3}{5}$.

Remember also that fractions can be converted to decimals. In some cases, it may be easier to work with decimals. Table 5–2 gives the decimal equivalents of common fractions.

TABLE 5–2

DECIMAL EQUIVALENTS OF FRACTIONS

Fraction	Decimal	Fraction	Decimal
$\frac{1}{64}$	.015625	$\frac{7}{16}$	.4375
$\frac{1}{32}$	.03125	$\frac{29}{64}$	.453125
$\frac{3}{64}$	.046875	$\frac{15}{32}$	.46875
$\frac{1}{20}$	.05	$\frac{31}{64}$	.484375
$\frac{1}{16}$	.0625	$\frac{1}{2}$	.5
$\frac{1}{13}$	.0769	$\frac{33}{64}$	.515625
$\frac{5}{64}$	.078125	$\frac{17}{32}$	.53125
$\frac{1}{12}$	.0833	$\frac{35}{64}$	.546875
$\frac{1}{11}$	.0909	$\frac{9}{16}$	.5625
$\frac{3}{32}$	.09375	$\frac{37}{64}$	.578125
$\frac{1}{10}$	.10	$\frac{19}{32}$	.59375
$\frac{7}{64}$	.109375	$\frac{39}{64}$	.609375
$\frac{1}{9}$	.111	$\frac{5}{8}$	.625
$\frac{1}{8}$	.125	$\frac{41}{64}$	.640625
$\frac{9}{64}$	.140625	$\frac{21}{32}$	.65625
$\frac{1}{7}$	.1429	$\frac{43}{64}$	.671875
$\frac{5}{32}$	.15625	$\frac{11}{16}$	.6875
$\frac{1}{6}$	.1667	$\frac{45}{64}$	.703125
$\frac{11}{64}$	.171875	$\frac{23}{32}$	.71875
$\frac{3}{16}$	.1875	$\frac{47}{64}$	.734375
$\frac{1}{5}$	.2	$\frac{3}{4}$	.75
$\frac{13}{64}$	.203125	$\frac{49}{64}$	.765625
$\frac{7}{32}$	.21875	$\frac{25}{32}$	.78125
$\frac{15}{64}$	.234375	$\frac{51}{64}$	.796875
$\frac{1}{4}$	.25	$\frac{13}{16}$	.8125
$\frac{17}{64}$	.265625	$\frac{53}{64}$	.828125
$\frac{9}{32}$	.28125	$\frac{27}{32}$	.84375
$\frac{19}{64}$	.296875	$\frac{55}{64}$	.859375
$\frac{5}{16}$	.3125	$\frac{7}{8}$	.875
$\frac{21}{64}$	.328125	$\frac{57}{64}$	.890625
$\frac{1}{3}$	.333	$\frac{29}{32}$	.90625
$\frac{11}{32}$	.34375	$\frac{59}{64}$	.921875
$\frac{23}{64}$	.359375	$\frac{15}{16}$	.9375
$\frac{3}{8}$	.375	$\frac{61}{64}$	.953125
$\frac{25}{64}$	.390625	$\frac{31}{32}$	.96875
$\frac{13}{32}$	.40625	$\frac{63}{64}$	.984375
$\frac{27}{64}$	.421875	1	1.00000

DECIMALS AND PERCENTAGES

A decimal is a division of one. A decimal point followed by numerals indicates the part of one. The decimal 0.25, for example, is equal to $^{25}/_{100}$ths, or $^{1}/_{4}$, of 1. As stated above, fractions and decimals are often used interchangeably. Decimals are also used when working with percentages.

To change a percentage to a decimal, remove the percent sign, count two places to the left (adding a zero if necessary), and add the decimal point.

$$25\% = .25$$

$$37\% = .37$$

$$7\% = .07$$

Let's work through a few examples using decimals and percentages.

■ EXAMPLE 1:

If the total price of plumbing supplies is $345.00 and there is a 5 percent tax, how much do you have to pay?

Solution: First you must figure what the tax is—what is 5 percent of $345.00? To do this, convert 5 percent to a decimal and multiply.

$$\begin{array}{r} \$\,345 \\ \times\,.05 \\ \hline \$\;17.25 \end{array}$$

Then add the tax to the price in order to determine the total amount that must be paid.

$$\begin{array}{r} \$\;345.00 \\ +17.25 \\ \hline \$\;362.25 \end{array}$$

Answer: $362.25

■ EXAMPLE 2:

You are purchasing miscellaneous supplies from a local plumbing supply store. The total costs amount to $678.00. The owner of the store offers you a 15 percent discount. How much will you actually pay for the supplies?

Solution: Determine what the discount is. What is 15 percent of $678.00?

Convert 15 percent to a decimal and multiply.

$$\begin{array}{r} \$\ 678 \\ \times\ .15 \\ \hline 3390 \\ 678\ \ \\ \hline \$\ 101.70 \end{array}$$

Now that you know that the discount amounts to $101.70, subtract it from the total price to determine what you will actually pay.

$$\begin{array}{r} \$\ 678.00 \\ -\$\ 101.70 \\ \hline \$\ 576.30 \end{array}$$

Answer: You will pay $576.30.

Percentages are often used in figuring out all the costs of operating a business. (See Figure 5–1.)

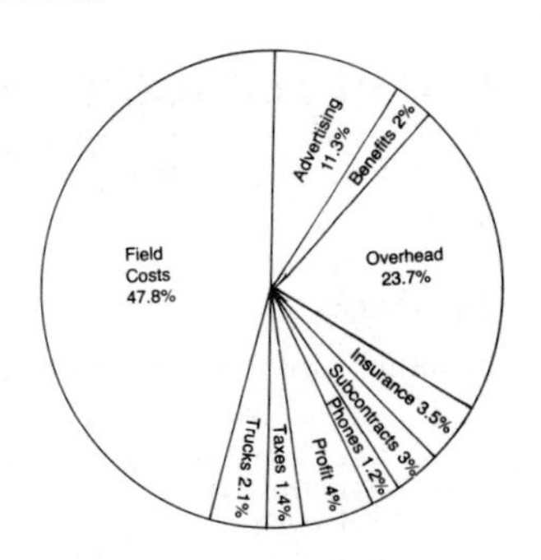

(Figure 5-1) Pie-chart showing the costs of operating a typical plumbing business

■ EXAMPLE 3:

If the owner of a plumbing service company expects to do a gross volume of business amounting to $800,000 and knows from previous experience that advertising costs usually run 11.3 percent, how much money should be allocated for advertising?

Solution: Convert 11.3 percent to a decimal and multiply the decimal by the gross amount.

$$
\begin{array}{r}
\$\ 800{,}000 \\
\times \qquad .113 \\
\hline
2400000 \\
800000 \\
800000 \\
\hline
\$\ 90{,}400.000
\end{array}
\qquad (11.3\%)
$$

Answer: $90,400 should be allocated for advertising.

GEOMETRY

Plumbers and apprentice plumbers often need to know the distance around an object, the total area of an object or location, and the volume (capacity) of a container. An understanding of the basic formulas for perimeter, area, and volume and how to use them is essential.

PERIMETER (CIRCUMFERENCE)

The perimeter is the distance around an object. In the case of a square, rectangle, or other polygon, it is the sum of the lengths of all sides.

RECTANGLE

The formula for the perimeter of a rectangle is

$$P = 2l + 2w$$

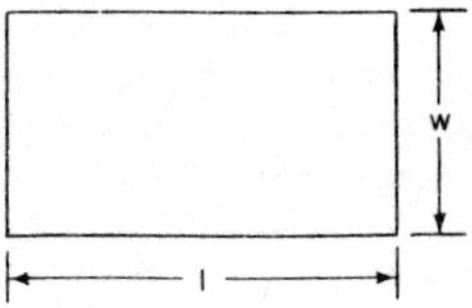

where P = perimeter
l = length
w = width

In this example, the perimeter is

w

l

$$P = 2l + 2w$$
$$P = 2 \times 9' + 2 \times 7'$$
$$P = 18' + 14'$$
$$P = 32'$$

SQUARE

The formula for the perimeter of a square is

$$P = 4s$$

where P = perimeter
s = side

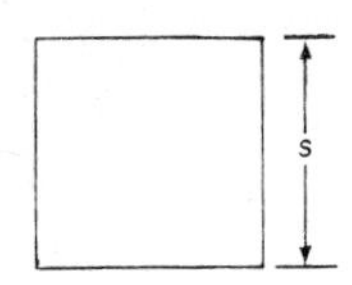

TRIANGLE

The formula for the perimeter of a triangle is

$$P = a + b + c$$

where P = perimeter
a = one side
b = second side
c = third side

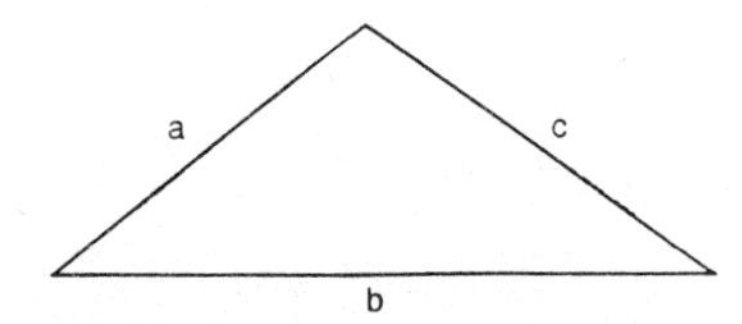

CIRCLE

The perimeter of a circle is known as the circumference of the circle. There are two formulas for the circumference of a circle, one using the diameter (d) of the circle, one using the radius (r) of the circle.

$$c = \pi d \quad \text{or} \quad c = 2\pi r$$

where c = circumference
π = pi, equal to 3.1416
d = diameter
r = radius

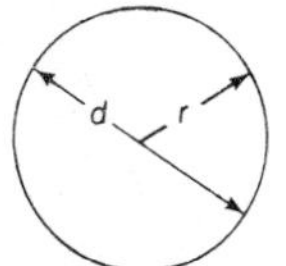

In the example given, the circumference can be calculated using either formula

$c = \pi d$
$c = 3.1416 \times 6''$
$c = 18.8496''$

$c = 2\pi r$
$c = 2 \times 3.1416 \times 3''$
$c = 18.8496''$

Note, in this case, the result should be rounded off to 18.85″.

Table 5–3 gives the circumference of circles with diameters from ⅛ to 99 (inch, foot, yard, meter, or other unit).

TABLE 5–3
CIRCUMFERENCE OF CIRCLES

Diameter	Circumference	Diameter	Circumference
1/8	.3927	10	31.41
1/4	.7854	10 1/2	32.98
3/8	1.178	11	34.55
1/2	1.570	11 1/2	36.12
5/8	1.963	12	37.69
3/4	2.356	12 1/2	39.27
7/8	2.748	13	40.84
1	3.141	13 1/2	42.41
1 1/8	3.534	14	43.98
1 1/4	3.927	14 1/2	45.55
1 3/8	4.319	15	47.12
1 1/2	4.712	15 1/2	48.69
1 5/8	5.105	16	50.26
1 3/4	5.497	16 1/2	51.83
1 7/8	5.890	17	53.40
2	6.283	17 1/2	54.97
2 1/4	7.068	18	56.54
2 1/2	7.854	18 1/2	58.11
2 3/4	8.639	19	59.69
3	9.424	19 1/2	61.26
3 1/4	10.21	20	62.83
3 1/2	10.99	20 1/2	64.40
3 3/4	11.78	21	65.97
4	12.56	21 1/2	67.54
4 1/2	14.13	22	69.11
5	15.70	22 1/2	70.68
5 1/2	17.27	23	72.25
6	18.84	23 1/2	73.82
6 1/2	20.42	24	75.39
7	21.99	24 1/2	76.96
7 1/2	23.56	25	78.54
8	25.13	26	81.68
8 1/2	26.70	27	84.82
9	28.27	28	87.96
9 1/2	29.84	29	91.10

TABLE 5–3—Continued

CIRCUMFERENCE OF CIRCLES

Diameter	Circumference	Diameter	Circumference
30	94.24	65	204.2
31	97.38	66	207.3
32	100.5	67	210.4
33	103.6	68	213.6
34	106.8	69	216.7
35	109.9	70	219.9
36	113.0	71	223.0
37	116.2	72	226.1
38	119.3	73	229.3
39	122.5	74	232.4
40	125.6	75	235.6
41	128.8	76	238.7
42	131.9	77	241.9
43	135.0	78	245.0
44	138.2	79	248.1
45	141.3	80	251.3
46	144.5	81	254.4
47	147.6	82	257.6
48	150.7	83	260.7
49	153.9	84	263.8
50	157.0	85	267.0
51	160.2	86	270.1
52	163.3	87	273.3
53	166.5	88	276.4
54	169.6	89	279.6
55	172.7	90	282.7
56	175.9	91	285.8
57	179.0	92	289.0
58	182.2	93	292.1
59	185.3	94	295.3
60	188.4	95	298.4
61	191.6	96	301.5
62	194.7	97	304.7
63	197.9	98	307.8
64	201.0	99	311.0

TABLE 5–4

SURFACE MEASURE	
144 sq. in.	= 1 sq. ft.
9 sq. ft.	= 1 sq. yd.
$30\frac{1}{4}$ sq. yd.	= 1 sq. rd.
160 sq. rd.	= 1 acre
640 acres	= 1 sq. mile
43,560 sq. ft.	= 1 acre

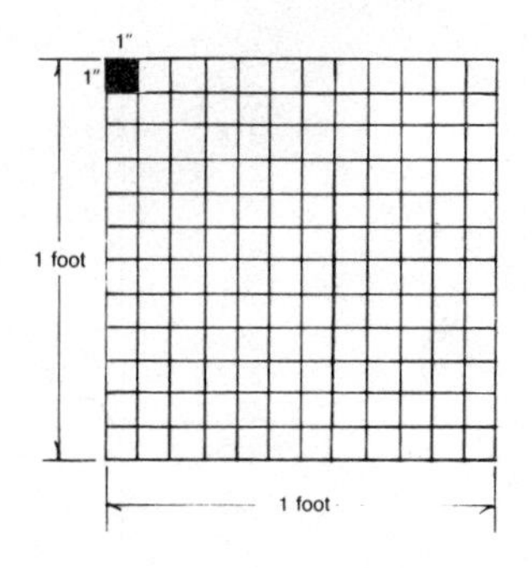

(Figure 5-2) 1 square yard = 144 square inches

AREA

Area is the number of unit squares that cover a figure. Remember that when you multiply linear units (for example, feet or inches or yards) by each other, the result—the product—is in square units (square feet, square inches, or square yards). (See Table 5–4 and Figure 5–2.)

RECTANGLE

The formula for the area of a rectangle is

$$A = l \times w$$

where A = area
l = length
w = width

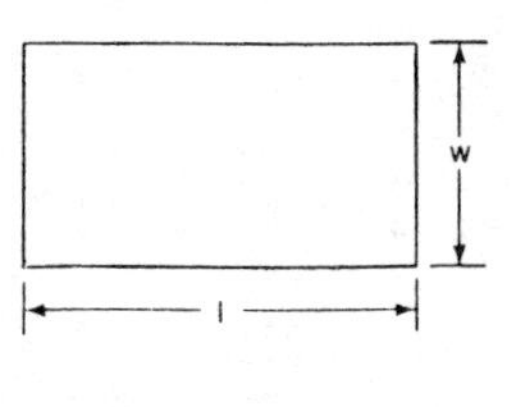

In the example given, the area is

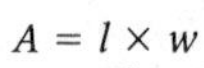

$$A = l \times w$$
$$A = 8' \times 3'$$
$$A = 24 \text{ square feet}$$

SQUARE

The formula for the area of a square is

$$A = s^2$$

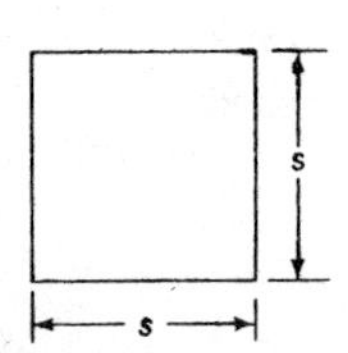

where A = area
s = side
^{2} = squared, or multiplied by itself

Since each side of a square is the same length, the area is obtained by multiplying $s \times s$, or s^2.

TRAPEZOID

The formula for the area of a trapezoid is

$$A = \frac{1}{2}(b_1 + b_2)\, h$$

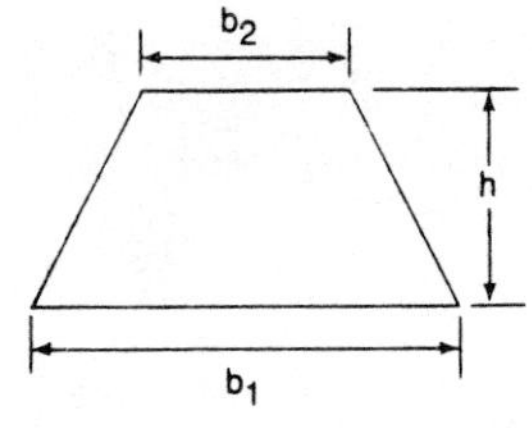

where A = area
b_1 = first base
b_2 = second base
h = height

Remember: the operation in the parenthesis must be performed first.

TRIANGLE

The formula for the area of a triangle is

$$A = \frac{1}{2}bh$$

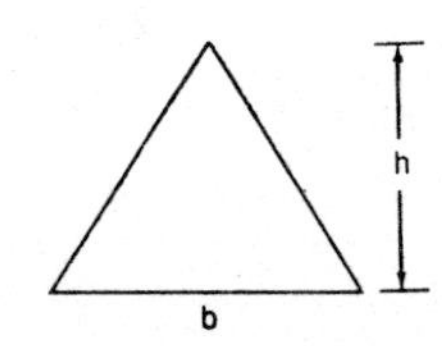

where A = area
b = base
h = height

CIRCLE

The formula for the area of a circle is

$$A = \pi r^2$$

where A = area
π = pi, equal to 3.1416
r = radius

In the example given,

$$A = \pi r^2$$
$$A = 3.1416 \times 9'' \times 9''$$
$$A = 3.1416 \times 81 \text{ square inches}$$
$$A = 254.4696 \text{ square inches, or } 254.47 \text{ square inches}$$

The area of a circle may also be calculated, using the following formula:

$$A = \frac{\pi}{4} d^2$$

In the example given above,

$$A = \frac{\pi}{4} d^2$$
$$A = \frac{3.1416}{4} \times 18^2$$
$$A = 0.7854 \times 324$$
$$A = 254.4696, \text{ or } 254.47 \text{ square inches}$$

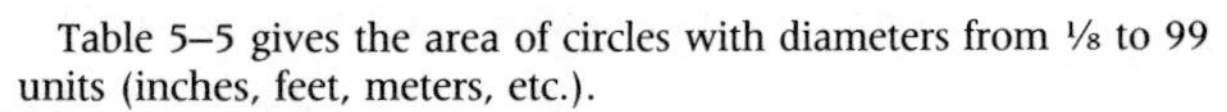

Table 5–5 gives the area of circles with diameters from ⅛ to 99 units (inches, feet, meters, etc.).

VOLUME

Volume is the number of unit cubes that fit inside an object. The volume of an object is given in cubic units (cubic feet, cubic yards, etc.). (See Table 5–6 and Figure 5–4.)

RECTANGLE

The formula for the volume of a rectangle

$$V = lwh \text{ or } V = l \times w \times h$$

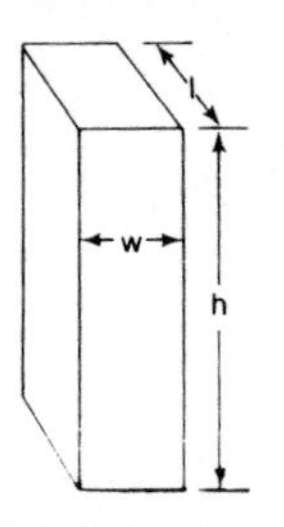

where V = volume
l = length
w = width
h = height (or depth)

In the example given, the volume is

$V = lwh$
$V = 6'' \times 4'' \times 15''$
$V = 360$ cubic inches

CUBE

The volume of a cube is

$$V = s^3$$

where V = volume
s = side
^{3} = cubed (in this case, $s \times s \times s$)

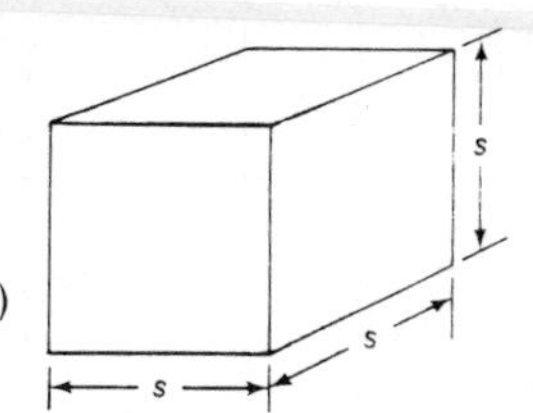

TRIANGULAR PRISM

The formula for the volume of a triangular prism is

$$V = \frac{abh}{2}$$

where V = volume
a = length of the prism
b = the base of the triangle
h = the height (depth) of the triangle

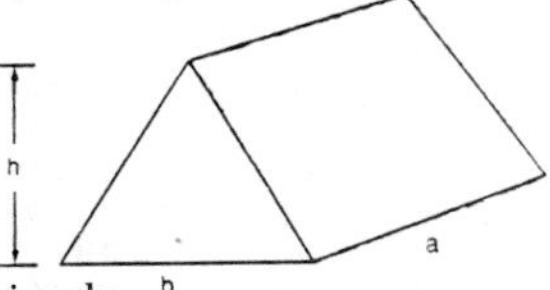

TRAPEZOIDAL PRISM

The volume of a trapezoidal prism is

$$V = \frac{(B + b)}{2} ah$$

where V = volume
B = larger base
a = length of the prism
b = smaller base
h = height (depth) of trapezoid

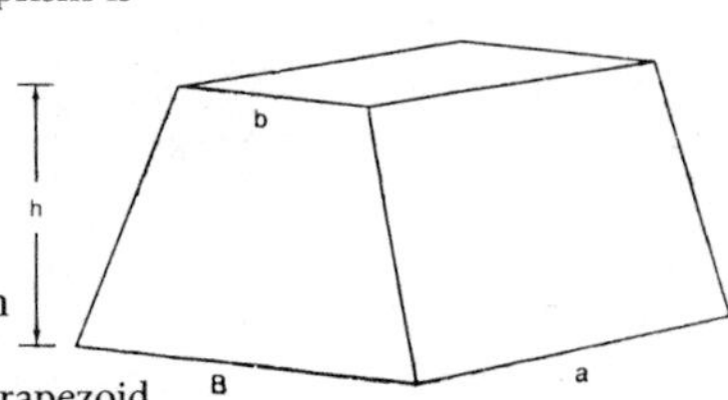

TABLE 5–5

AREA OF CIRCLES

Diameter	Area	Diameter	Area
$\frac{1}{8}$	0.0123	10	78.54
$\frac{1}{4}$	0.0491	$10\frac{1}{2}$	86.59
$\frac{3}{8}$	0.1104	11	95.03
$\frac{1}{2}$	0.1963	$11\frac{1}{2}$	103.86
$\frac{5}{8}$	0.3068	12	113.09
$\frac{3}{4}$	0.4418	$12\frac{1}{2}$	122.71
$\frac{7}{8}$	0.6013	13	132.73
1	0.7854	$13\frac{1}{2}$	143.13
$1\frac{1}{8}$	0.9940	14	153.93
$1\frac{1}{4}$	1.227	$14\frac{1}{2}$	165.13
$1\frac{3}{8}$	1.484	15	176.71
$1\frac{1}{2}$	1.767	$15\frac{1}{2}$	188.69
$1\frac{5}{8}$	2.073	16	201.06
$1\frac{3}{4}$	2.405	$16\frac{1}{2}$	213.82
$1\frac{7}{8}$	2.761	17	226.98
2	3.141	$17\frac{1}{2}$	240.52
$2\frac{1}{4}$	3.976	18	254.46
$2\frac{1}{2}$	4.908	$18\frac{1}{2}$	268.80
$2\frac{3}{4}$	5.939	19	283.52
3	7.068	$19\frac{1}{2}$	298.6
$3\frac{1}{4}$	8.295	20	314.16
$3\frac{1}{2}$	9.621	$20\frac{1}{2}$	330.06
$3\frac{3}{4}$	11.044	21	346.36
4	12.566	$21\frac{1}{2}$	363.05
$4\frac{1}{2}$	15.904	22	380.13
5	19.635	$22\frac{1}{2}$	397.60
$5\frac{1}{2}$	23.758	23	415.47
6	28.274	$23\frac{1}{2}$	433.73
$6\frac{1}{2}$	33.183	24	452.39
7	38.484	$24\frac{1}{2}$	471.43
$7\frac{1}{2}$	44.178	25	490.87
8	50.265	26	530.93
$8\frac{1}{2}$	56.745	27	572.55
9	63.617	28	615.75
$9\frac{1}{2}$	70.882	29	660.52

TABLE 5–5—Continued

AREA OF CIRCLES

Diameter	Area	Diameter	Area
30	706.86	65	3318.3
31	754.76	66	3421.2
32	804.24	67	3525.6
33	855.30	68	3631.6
34	907.92	69	3739.2
35	962.11	70	3848.4
36	1017.8	71	3959.2
37	1075.2	72	4071.5
38	1134.1	73	4185.4
39	1194.5	74	4300.8
40	1256.6	75	4417.8
41	1320.2	76	4536.4
42	1385.4	77	4656.6
43	1452.2	78	4778.3
44	1520.5	79	4901.6
45	1590.4	80	5026.5
46	1661.9	81	5153.0
47	1734.9	82	5281.0
48	1809.5	83	5410.6
49	1885.7	84	5541.7
50	1963.5	85	5674.5
51	2042.8	86	5808.8
52	2123.7	87	5944.6
53	2206.1	88	6082.1
54	2290.2	89	6221.1
55	2375.8	90	6361.7
56	2463.0	91	6503.9
57	2551.7	92	6647.6
58	2642.0	93	6792.9
59	2733.9	94	6939.8
60	2827.4	95	7088.2
61	2922.4	96	7238.2
62	3019.0	97	7389.8
63	3117.2	98	7542.9
64	3216.9	99	7697.7

TABLE 5–6

CUBIC MEASURE	
1728 cu. in.	= 1 cu. ft.
27 cu. ft.	= 1 cu. yd.
128 cu. ft.	= 1 cord

CYLINDER

A plumber frequently needs to calculate the capacity, or volume, of a cylinder. The formula for the volume of a cylinder is

$$V = \pi r^2 h$$

where V = volume
r = radius
h = height
π = pi, equal to 3.1416

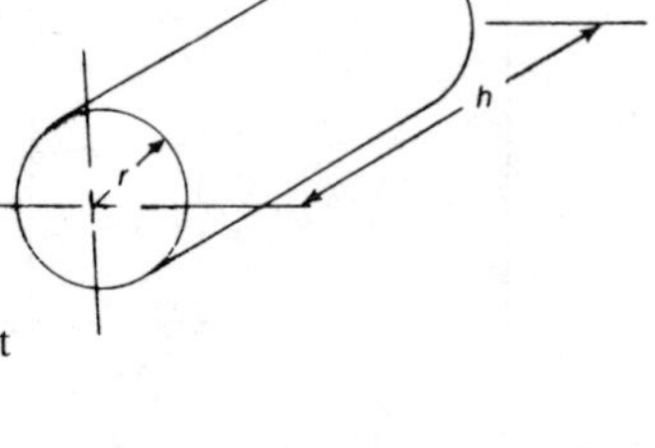

In the example given,

$$V = \pi r^2 h$$
$$V = 3.1416 \times 5' \times 5' \times 8'$$
$$V = 3.1416 \times 25 \text{ square feet} \times 8 \text{ feet}$$
$$V = 3.1416 \times 200 \text{ cubic feet}$$
$$V = 628.32 \text{ cubic feet}$$

There is another formula that can be used to determine the volume of a cylinder

$$V = \frac{\pi}{4} d^2 h$$

where V = volume
d = diameter
h = height
π = pi, equal to 3.1416

TRIGONOMETRY

Trigonometry is used when working with triangles. There are six basic relationships that express the relationship of the sides of a right triangle. These relationships, used with the table of trigonometric functions (see Table 5–7), are used to find the sides and angles of a triangle. In plumbing situations, trigonometry is used to determine slopes, slanting surfaces, angles, etc.

The six basic relationships are:

Sine

$$\sin = \frac{\text{side opposite}}{\text{hypotenuse}}$$

Cosine

$$\cos = \frac{\text{side adjacent}}{\text{hypotenuse}}$$

Tangent

$$\tan = \frac{\text{side opposite}}{\text{side adjacent}}$$

Cosecant

$$\csc = \frac{\text{hypotenuse}}{\text{side opposite}}$$

Secant

$$\sec = \frac{\text{hypotenuse}}{\text{side adjacent}}$$

Cotangent

$$\cot = \frac{\text{side adjacent}}{\text{side opposite}}$$

FORMULAS OFTEN USED BY PLUMBERS

In addition to the basic formulas presented in the previous sections of this chapter, the plumber often uses other formulas to calculate grade, run, pitch, offsets, fittings, and other factors needed in daily work. We shall briefly present some of these formulas and explain how to use them.

GALLONS IN A TANK

A plumber often needs to know how many gallons a particular tank can hold. If the diameter (d) of the tank and the height or length (h) of the tank are known, the gallon capacity of the tank can be determined. (See Figure 5–3.)

TABLE 5–7
TRIGONOMETRIC FUNCTIONS

Angle	Sine	Cosine	Tangent	Angle	Sine	Cosine	Tangent
1°	0.0175	0.9998	0.0175	21°	0.3584	0.9336	0.3839
2°	0.0349	0.9994	0.0349	22°	0.3746	0.9272	0.4040
3°	0.0523	0.9986	0.0524	23°	0.3907	0.9205	0.4245
4°	0.0698	0.9976	0.0699	24°	0.4067	0.9135	0.4452
5°	0.0872	0.9962	0.0875	25°	0.4226	0.9063	0.4663
6°	0.1045	0.9945	0.1051	26°	0.4384	0.8988	0.4877
7°	0.1219	0.9925	0.1228	27°	0.4540	0.8910	0.5095
8°	0.1392	0.9903	0.1405	28°	0.4695	0.8829	0.5317
9°	0.1564	0.9877	0.1584	29°	0.4848	0.8746	0.5543
10°	0.1736	0.9848	0.1763	30°	0.5000	0.8660	0.5774
11°	0.1908	0.9816	0.1944	31°	0.5150	0.8572	0.6009
12°	0.2079	0.9871	0.2126	32°	0.5299	0.8480	0.6249
13°	0.2250	0.9744	0.2309	33°	0.5446	0.8387	0.6494
14°	0.2419	0.9703	0.2493	34°	0.5592	0.8290	0.6745
15°	0.2588	0.9659	0.2679	35°	0.5736	0.8192	0.7002
16°	0.2756	0.9613	0.2867	36°	0.5878	0.8090	0.7265
17°	0.2924	0.9563	0.3057	37°	0.6018	0.7986	0.7536
18°	0.3090	0.9511	0.3249	38°	0.6157	0.7880	0.7813
19°	0.3256	0.9455	0.3443	39°	0.6293	0.7771	0.8098
20°	0.3420	0.9397	0.3640	40°	0.6428	0.7660	0.8391

41°	0.6561	0.7547	0.8693	66°	0.9135	0.4067	2.2460
42°	0.6691	0.7431	0.9004	67°	0.9205	0.3907	2.3559
43°	0.6820	0.7314	0.9325	68°	0.9272	0.3746	2.4751
44°	0.6947	0.7193	0.9657	69°	0.9336	0.3584	2.6051
45°	0.7071	0.7071	1.0000	70°	0.9397	0.3420	2.7475
46°	0.7193	0.6947	1.0355	71°	0.9455	0.3256	2.9042
47°	0.7314	0.6820	1.0724	72°	0.9511	0.3090	3.0777
48°	0.7431	0.6691	1.1106	73°	0.9563	0.2924	3.2709
49°	0.7547	0.6561	1.1504	74°	0.9613	0.2756	3.4874
50°	0.7660	0.6428	1.1918	75°	0.9659	0.2588	3.7321
51°	0.7771	0.6293	1.2349	76°	0.9703	0.2419	4.0108
52°	0.7880	0.6157	1.2799	77°	0.9744	0.2250	4.3315
53°	0.7986	0.6018	0.3270	78°	0.9781	0.2079	4.7046
54°	0.8090	0.5878	1.3764	79°	0.9816	0.1908	5.1446
55	0.8192	0.5736	1.4281	80°	0.9848	0.1736	5.6713
56°	0.8290	0.5592	1.4826	81°	0.9877	0.1564	6.3138
57°	0.8387	0.5446	1.5399	82°	0.9903	0.1392	7.1154
58°	0.8480	0.5299	1.6003	83°	0.9925	0.1219	8.1443
59°	0.8572	0.5150	1.6643	84°	0.9945	0.1045	9.5144
60°	0.8660	0.5000	1.7321	85°	0.9962	0.0872	11.4301
61°	0.8746	0.4848	1.8040	86°	0.9976	0.0698	14.3007
62°	0.8829	0.4695	1.8807	87°	0.9986	0.0523	19.0811
63°	0.8910	0.4540	1.9626	88°	0.9994	0.0349	28.6363
64°	0.8988	0.4384	2.0503	89°	0.9998	0.0175	57.2900
65°	0.9063	0.4226	2.1445	90°	1.0000	0.0000	

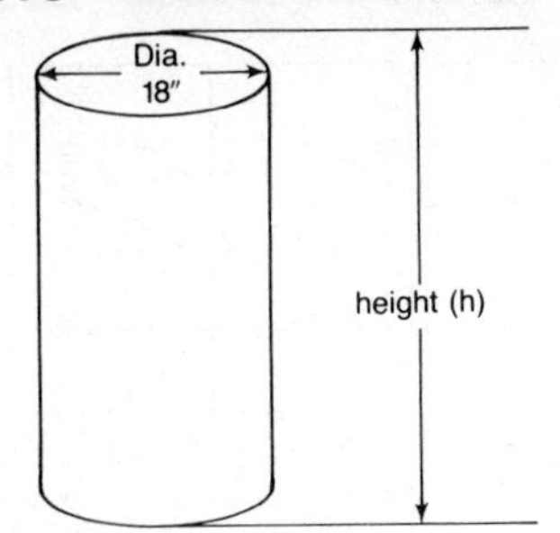

(Figure 5-3) The capacity in gallons of a tank can be determined, if its height and diameter are known.

If the diameter and height are measured in inches, the formula is

$$G = d^2 \times h \times 0.0034$$

If the diameter and height are measured in feet, the formula is

$$G = d^2 \times h \times 7.5$$

■ EXAMPLE 1:

If a tank has a diameter of 18 inches and a height of 60 inches, how many gallons can it hold?

Solution: Choose the proper formula—in this case,

$$G = d^2 \times h \times 0.0034$$

Substitute the known quantities in the formula,

$$G = 18^2 \times 60 \times 0.0034$$

Do the arithmetic.

$$G = 324 \times 60 \times 0.0034$$
$$G = 19440 \times 0.0034$$
$$G = 66.096^*$$

Answer: 66 gallons

Note: In this example, the answer would be 66 gallons. You would *not* round off to 66.1 gallons. (The exact figure is a maximum, so you cannot round off upwards.)

■ EXAMPLE 2:

If a tank has a height of 7 feet and a diameter of 3 feet, how many gallons can it hold?

Solution: Choose the proper formula—in this case,

$$G = d^2 \times 0.7854 \times h \times 7.5$$

Substitute the known quantities in the formula

$$G = 3^2 \times 0.7854 \times 7 \times 7.5$$

Do the arithmetic

$$G = 7.0686 \times 52.5$$
$$G = 371.1015 \text{ gallons}$$

Answer: 371 gallons

GRADE, RUN, AND DROP FORMULA

When installing sanitary sewer lines, the plumber must calculate the length of the run, the drop necessary, and the grade, or pitch, necessary to ensure proper flow through the pipe. This is especially true when the pipes must be run into a sewer connection already existing on the street.

Most plumbing installations are graded—usually ¼ inch per foot—to allow wastewater and waste products to flow smoothly through waste piping. (In some cases ⅛ inch per foot pitch is allowed by code.)

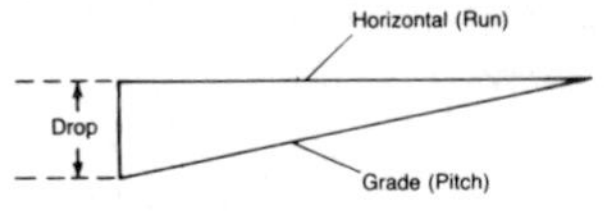

(Figure 5-4) Grade, run, and drop

Before discussing the formula that plumbers can use to determine grade, run, or drop, let's define the terms. (See Figure 5–4).

Run (R) is the horizontal distance measured in feet and/or inches.
Drop (D) is the amount down from level or the horizontal run measured in feet and/or inches.
Grade (G), or *pitch,* is the slope of the pipe in inches per foot.

Run, drop, and grade are related according to the following formula:

$$D = G \times R$$

This formula can also be expressed as

$$G = \frac{D}{R} \qquad \text{or} \qquad R = \frac{D}{G}$$

■ EXAMPLE 1:

If the horizontal run is known to be 60 feet and the drop is 15 inches, what grade, or pitch, must be used for the pipe?

Solution: First, determine which expression of the formula to use. Since you know the horizontal run (R) and the drop (D), use

$$G = \frac{D\text{ (number of inches)}}{R\text{ (number of feet)}}$$

Substitute the known factors.

$$G = \frac{15}{60}$$

Do the math.

$$G = 60\overline{)15.00}\quad .25$$

Answer: .25 inches per foot, or ¼ inch per foot, is the pitch required.

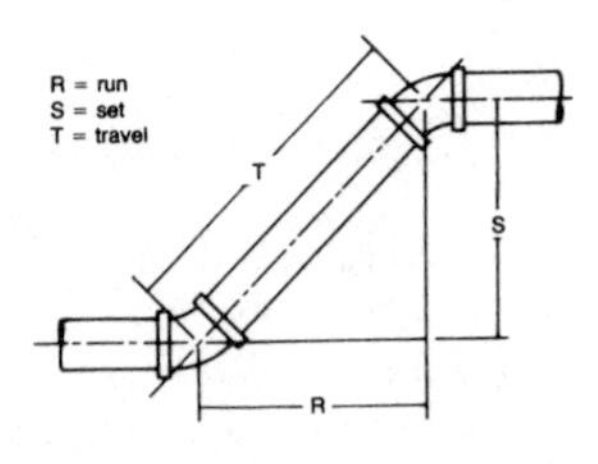

(Figure 5-5) Terms used to describe an offset

OFFSETS

The plumber must frequently calculate the length of pipe needed when changing directions. Tables of constants and formulas for different types of commonly used fittings are available to enable the plumber to do this.

Figure 5–5 shows clearly the terms used for various parts of an offset.

Table 5–8 gives the formula and constants for calculating parts of an offset when using any of six different fittings.

■ EXAMPLE 1:

If the set is known to be 15 inches, and you are using a 45 degree ell fitting, what is the travel? In other words, how much pipe is needed on the diagonal?

Solution: Refer to the appropriate table—in this case, Table 5–8. Since you know the set (S) and wish to find the travel

TABLE 5–8

CONSTANTS FOR MEASURING OFFSETS IN PIPING

known side	to find side	multiply side	using $5\frac{5}{8}$ ell	using $11\frac{1}{4}$ ell	using $22\frac{1}{2}$ ell	using 30 ell	using 45 ell	using 60 ell
S	*T*	*S*	10.19	5.13	2.61	2.00	1.41	1.15
S	*R*	*S*	10.16	5.03	2.41	1.73	1.00	.58
R	*S*	*R*	.10	.20	.41	.58	1.00	1.73
R	*T*	*R*	1.00	1.02	1.08	1.16	1.41	2.00
T	*S*	*T*	.10	.20	.38	.50	.71	.87
T	*R*	*T*	1.00	.98	.92	.87	.71	.50

(*T*), use the first horizontal line of the table.
Find the column for the 45 degree ell.
Use the formula given and substitute.

$$T = S \times 1.41$$
$$T = 15'' \times 1.41$$
$$T = 21.15''$$

Answer: The travel is 21.15 inches, or, in other words, 21.15 inches of pipe are needed.

Remember, if you are using a U.S. Customary ruler to measure the pipe, you will need to consult a decimal equivalent chart (Table 5–2) to convert .15 inches to a fraction that you can read on your ruler. Looking at the numbers in the decimal column, you see that .15625 equals 5/32. Thus, you would measure and cut 21 5/32 inches of pipe.

Other tables and formulas are available for many types of fittings. In general, if one side of an offset and the type of fitting is known, the other parts of the offset can be calculated. Tables 5–9, and 5–10 give constants for parallel and rolling offsets. Table 5–11 gives further data, providing illustrations and formulas to help in calculating offsets with several other types of fittings.

TABLE 5–9
CONSTANTS FOR PARALLEL OFFSETS

Fitting Angle	90°	72°	60°	45°	22½°	11¼°
Diagonal = Offset ×	—	1.052	1.154	1.414	2.613	5.126
Rise (Run) = Offset ×	—	.325	.577	1.	2.414	5.027
Parallel Angle	45°	36°	30°	22½°	11¼°	5⅝°
Difference in Length = Spread ×	1.	.727	.577	.414	.199	.098

TABLE 5–10
CONSTANTS FOR ROLLING OFFSETS

Fitting Angle	90°	72°	60°	45°	22½°	11¼°
Diagonal	1	1.052	1.154	1.414	2.613	5.126
Setback	0	.325	.577	1.	2.414	5.027

TABLE 5–11
Constants for 45° Fittings

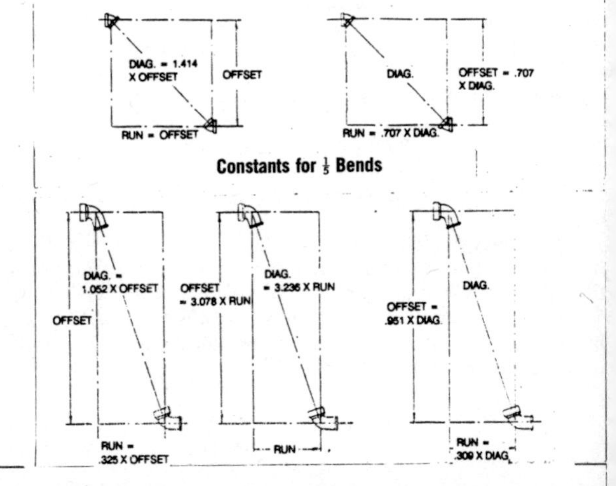

Constants for 60° Fittings

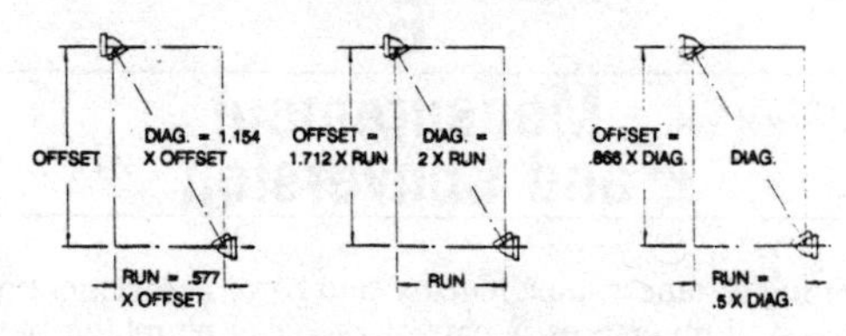

Constants for $22\frac{1}{2}$° Fittings

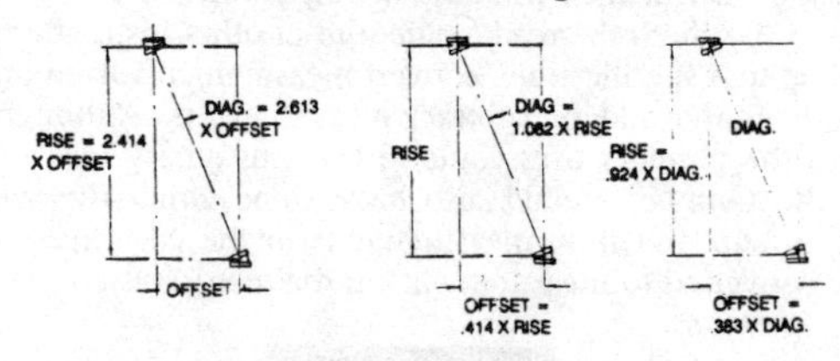

Constants for $11\frac{1}{4}$° Fittings

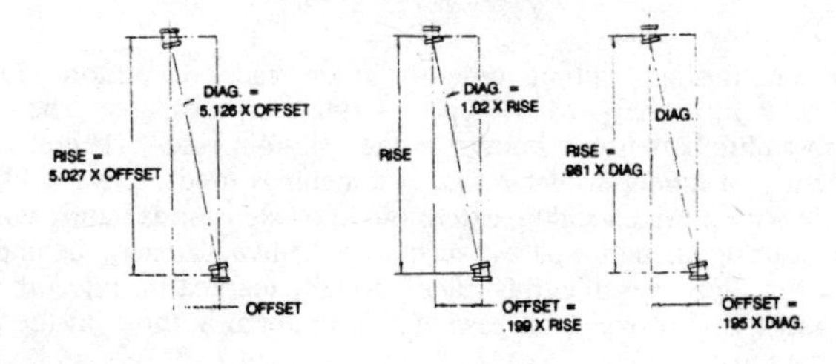

6

Measurement and Conversion

Measuring distances, pipe lengths, and correct placement of pipes, fixtures, and appliances is part of everyday plumbing work. The plumber—and apprentice—must know not only how to read measuring instruments, primarily a ruler, but also how to use a ruler to a specific scale, read a blueprint or other instructions done according to a specific scale, convert measurements from one unit to another, and add or subtract measurements. Although most supplies the plumber uses come in U.S. customary measurement units, the plumber should also have some familiarity with the metric system and how measurements in the customary system can be converted to measurements in the metric system.

USING A RULER

No one in the plumbing trade—or in any trade—can afford to use a ruler incorrectly. Many types of rulers are available. The one most plumbers use is known as the "six-foot rule." (There is an eight-foot model available that is sometimes used). It is a folding rule with a 6-inch sliding extension. The extension is handy when measuring in tight spaces, such as a hallway area or behind a fixture. Very versatile, this ruler is usually marked in sixteenths of an inch and provides for ease of measurement in most on-the-job situations.

Figure 6–1 shows an enlarged inch and its sixteenth divisions. Figure 6–2 shows how to read such a ruler. Study it and be sure to learn how to read it accurately and quickly. Notice that many of the fractions can be reduced to lower terms—for example, 6/16 equals 3/8 and 10/16 equals 5/8.

Fractional measurements may be converted to decimals, and, in some cases, you may find that working (adding or subtracting) with decimals is more convenient than working with fractions.

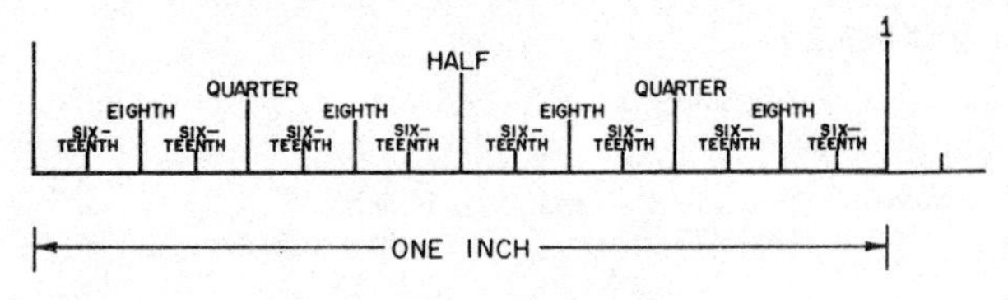

(Figure 6-1) The fractional divisions of an inch ruler

READING A RULER ACCORDING TO A SCALE

Most architects and engineers use a scale when making blueprints or other drawings. Otherwise, the drawing itself would be too large and inconvenient to use. The plumber must be able to read and interpret the scale if he or she wants to understand the sizes and dimensions called for and be able to complete the job as required. An architect's scale ruler is especially accurate and should be used in some situations.

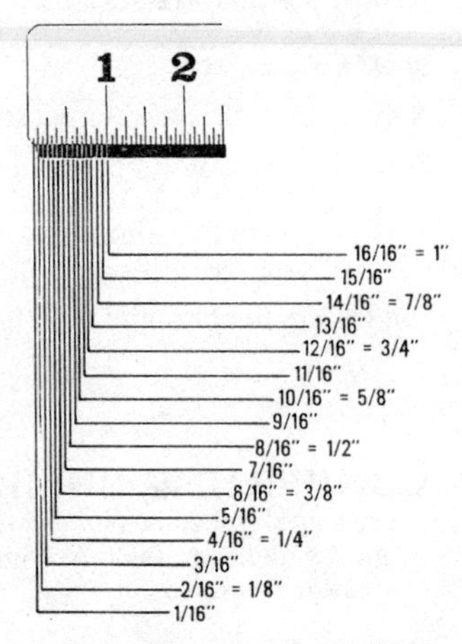

Figure 6-2) How to read a ruler marked in sixteenths

The scale an architect or engineer uses on a blueprint is noted directly on the blueprint, often in the title block. The scale may be given as ⅛″ = 1′ (⅛ inch on the ruler equals 1 foot) or it may be given as

1″ = 50′	2″ = 100′
¼″ = 50′	2″ = 400′
⅛″ = 50′	2″ = 800′

A few examples will help clarify the process of reading to scale.

■ EXAMPLE 1:

If the blueprint states that ⅛″ = 1′, how many feet are indicated by 3″?

Solution: There are 8 ⅛″ marks in 1 inch.
If each ⅛″ mark equals 1′, then 8 marks equal 8′.
Then 16 marks in 2″ equals 16 ′ and 24 marks in 3″ equals 24′.

Answer: 3 inches signifies 24 feet.

■ EXAMPLE 2:

If the scale on the blueprint is 1⁄16″ equals 1′, what does 3″ signify?

Solution: There are 16 marks in 1″.
If each mark equals 1′, then 1″ represents 16′.
Then 2″ represents 32′, and 3″ represents 48′.
(or, $16 \times 3 = 48$)

Answer: On this blueprint, 3 inches represents 48 feet.

CONVERTING UNLIKE MEASUREMENTS

When using a ruler, the plumber must often change feet into inches or inches into feet. Table 6–1 gives the equivalents for standard U.S. linear measurement units.

■ EXAMPLE 1:

How many inches are there in 12 feet?

Solution: 12×12

Answer: 144 inches

■ EXAMPLE: 2

How many feet are there in 48 inches?

Solution: $48 \div 12$

Answer: 4 feet

TABLE 6–1

LINEAR MEASURE	
12 inches (in.)	= 1 foot (ft.)
3 ft.	= 1 yard (yd.)
$16\frac{1}{2}$ ft.	= 1 rod (rd.)
$5\frac{1}{2}$ yd.	= 1 rd.
320 rd.	= 1 mile
1760 yd.	= 1 mile
5280 ft.	= 1 mile

ADDING MEASUREMENTS

If measurements are given in the same base unit—for example, inches or feet—they can be simply added or subtracted. If, however, they are given in different units—for example, inches and feet—each measurement must be converted to the same base unit before they can be added or subtracted.

If the measurements are in fractional units, be sure that the denominators of the fractions are the same before adding or subtracting. (Review the section on adding and subtracting fractions in Chapter 5.)

■ EXAMPLE:

Add 3′9″ and 42″

Solution: Change the measurements to the same base unit.

$$3'9'' = (3 \times 12'') + 9'' = 45''$$

Add

$$\begin{array}{r} 45'' \\ \underline{42''} \\ 87'' \end{array}$$

Answer: 87 inches, or 7 feet, 3 inches

$$87'' \div 12'' = 7'3''$$

■ **EXAMPLE 2:**

What is the total length of piping needed for three straight runs that measure 3 feet, 2 inches; 4 feet; and 6 inches?

Solution: First convert each measurement to the same unit and then add

3 feet, 2 inches	=	(3 × 12) + 2	= 38
4 feet	=	(4 × 12)	= 48
6 inches			= 6
			92 inches

Answer: 92 inches, or 7 feet, 8 inches

(92 ÷ 12 = 7 feet, 8 inches)

DETAILS SPECIFIC TO PIPE AND FITTING MEASUREMENT

The plumber and apprentice must also be aware of a few conventions that apply specifically to measuring pipes and fittings.

Copper fittings and threaded (for example, brass) pipe fittings are not standardized. (Iron pipe size fittings are.) Therefore, the plumber must measure the individual fitting on the job site.

There are several ways pipe and fittings can be measured. The way the measurement is taken must be stated along with the measured distance to ensure accuracy and the proper fitting. Figure 6–3 shows clearly how pipe and fittings can be measured.

METRIC SYSTEM

Although virtually all plumbing supplies in the United States (e.g., pipes, fittings) are specified according to the U.S. customary system of measurement (inches, feet etc.), the plumber and apprentice

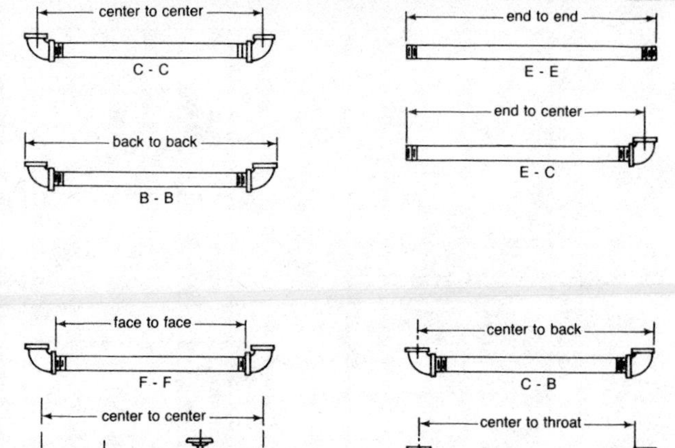

(Figure 6-3) Ways of measuring pipe and fitting allowances

should have a basic understanding of the metric system of measurement. In 1975 the Metric Conversion Act was signed into law, committing the United States to a voluntary conversion of all measurements to the metric system.

The metric system is more correctly known as the Système International d'Unités, which is abbreviated SI. Measurements in the United States (British) customary system can be converted to metric measurements—and vice versa—using standard conversion factors. Tables 6–2, 6–3, and 6–4 give the conversion factors for length units, area measurements, and volume measurements.

Using these tables is quite simple. To convert from one measurement unit to another, simply find the original unit listed in the left column. Then look across the table to the column headed by the unit you wish to convert to. Then simply multiply the original value by the conversion factor given. For example, to convert 6 feet to meters, use Table 6–2, find feet listed in the left column and follow that line across to the meter column. Then multiply 6 by

TABLE 6–2
LENGTH CONVERSION FACTORS

	inch	feet	yard	mile	millimeter	centimeter	meter	kilometer
1 in.	1	0.0833	0.0278	...	25.40	2.540	0.0254	...
1 ft.	12	1	0.333	...	304.8	30.48	0.3048	...
1 yd.	36	3	1	...	914.4	91.44	0.9144	...
1 mile	...	5280	1760	1	...	...	1609.3	1.609
1 mm	0.0394	0.0033	...	...	1	0.100	0.001	...
1 cm	0.3937	0.0328	0.0109	...	10	1	0.01	...
1m	...	3.281	1.094	...	1000	100	1	0.001
1 km		3281	1094	0.6214	...	...	1000	1

TABLE 6–3
AREA CONVERSION FACTORS

	square inch	square feet	acre	square mile	square centimeter	square meter
1 sq. in.	1	0.0069	...	...	6.452	...
1 sq. ft.	144	1	...	...	929.0	0.0929
1 acre	...	43,560	1	0.0016	...	4047
1 sq. mile	...	...	640	1	...	...
1 sq. cm	0.1550	...	...	...	1	0.0001
1 sq. m	1550	10.76	...	...	10,000	1

TABLE 6–4
VOLUME CONVERSION FACTORS

	cubic inch	cubic foot	cubic yard	cubic centimeter	cubic meter	liter	U.S. gallon
1 cu. in.	1	. . .	. . .	16.387	. . .	0.0164	. . .
1 cu. ft.	1728	1	0.0370	28,317	0.0283	28.32	7.481
1 cu yd.	46,656	27	1	. . .	0.7646	764.5	202.0
1 cu. cm	0.0610	. . .	. . .	1	. . .	0.0010	. . .
1 cu. m	61,023	35.31	1.308	1,000,000	1	999.97	264.2
1 liter	61,025	0.0353	. . .	1,000.028	0.0010	1	0.2642
1 US gal.	231	0.1337	. . .	3785.4	. . .	3.785	1
1 Imp. gal.	277.4	0.1605	. . .	4546.1	. . .	4.546	1.201

0.3048. The answer is 1.8288 meters. In other words, 6 feet equals 1.8 meters.

Appendix 3 provides several tables listing English–SI length, area, and volume equivalents as well as other tables providing information of interest to plumbers, such as flow rate conversions.

7

How To Read Blueprints

A plumber must be able to read blueprints in order to understand what is to be done and where it is to be done at a building site, to make estimates when bidding for a job, and to order supplies for a job.

Blueprints are a type of graphic language that permit a large amount of information to be presented in a clear and condensed form that is easy to understand and is conveniently placed on one (or a few) drawings that can be easily carried to the job site and/or stored. The language is made up of symbols, lines, abbreviations, and various types of drawings that are standard throughout the plumbing trade in the United States. Every plumber and apprentice plumber should be familiar with these symbols.

This chapter presents a brief look at the symbols and abbreviations and other conventions used on blueprints that a plumber is most likely to use.

PLOT PLAN

The first step in the development of a complete set of blueprints for a building is a plot plan (site plan). A plot plan shows the relationship between the building and the property, or plot, on which it is to be built. Only an outline of the building is given, but details about the site are included.

A plot plan shows an overview of the property lines, including street names, and all adjoining property lines. Existing structures on the plot are shown. Also included are all waste handling piping or septic system facilities; water mains; sewers; all water lines and waste lines; electrical and telephone lines; and all other information pertinent to the building site.

A plot plan, like most blueprints, is drawn to scale, and the scale is noted on the drawing itself. A scale of ¼″ = 1′ or 1″ = 50′ is not uncommon. (Refer to Chapter 6 for a discussion of reading to scale.)

Figure 7–1 shows a typical plot plan.

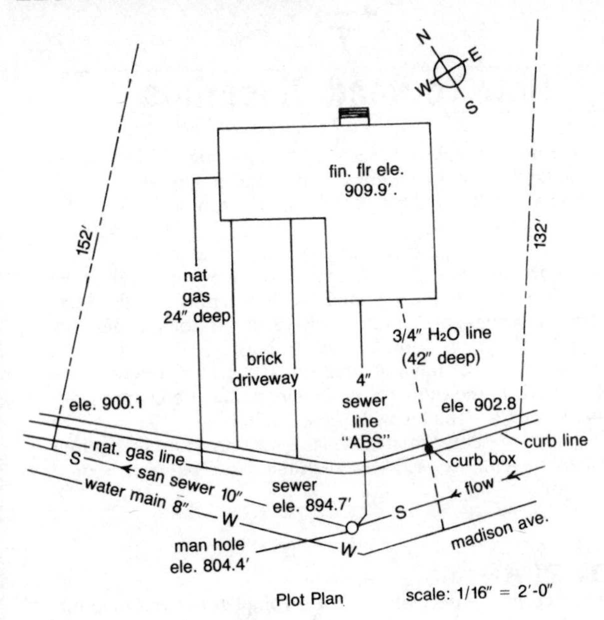

(Figure 7-1) Typical plot, or site, plan

ARCHITECTURAL AND STRUCTURAL DRAWINGS

After the plot plan come various stages of architectural and structural drawings. These drawings provide information on building foundations, footings, framing requirements, and roof details as well as plans for each floor of the building. Professionals in the building industries should be able to understand these plans to be able to determine if there is any information or factor pertinent to his/her particular part of the construction. For example, does the nature of the subsoil or the foundation influence the type of piping to be used? Figure 7–2 shows the symbols used on architectural drawings.

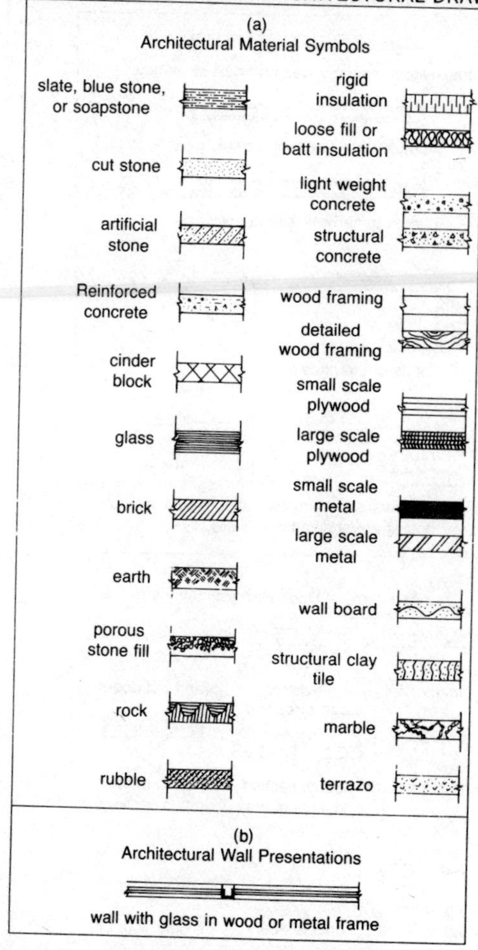

(Figure 7-2a) Symbols used in architectural drawings

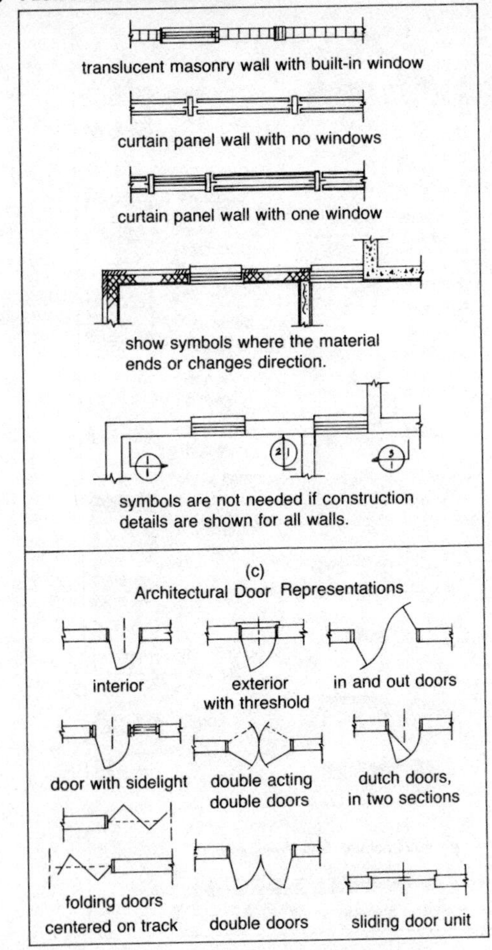

(Figure 7-2b) Symbols used in architectural drawings

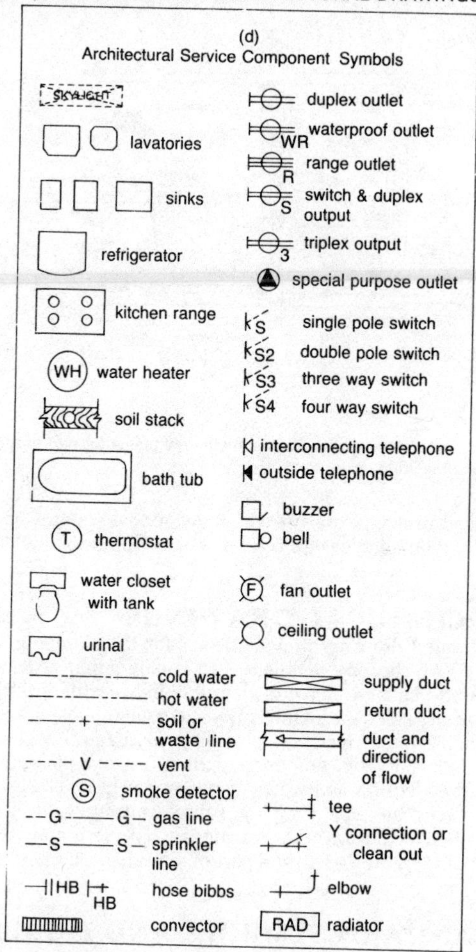

(Figure 7-2c) Symbols used in architectural drawings

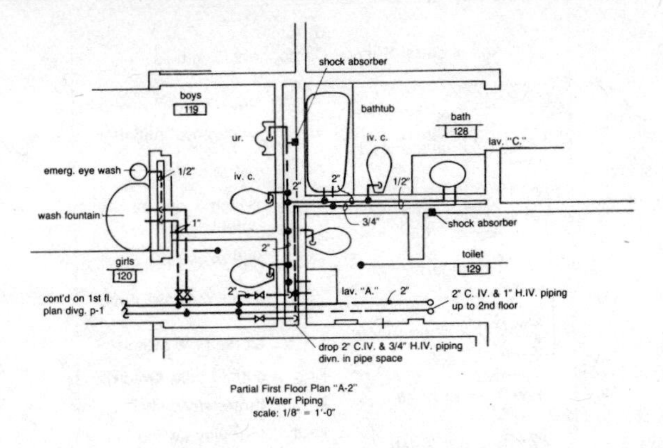

(Figure 7-3) Mechanical plans showing the piping for part of the first floor of a large building

Detail drawings are used to show special features of a certain part of a building or of a certain system within the building.

MECHANICAL DRAWINGS

Mechanical drawings provide details for the plumbing, electrical, and HVAC (heating–ventilating–air conditioning) systems. (If mechanical plans are not provided, a plumber with special training will need to take the architectural and structural plans and develop mechanical drawings.) A plumbing mechanical will show the design, layout, and component parts of the total plumbing system, including fixtures and water, waste, and vent lines. Specifications concerning the type of piping, valves, and fittings to be used are also often included. These drawings are schematic and use standard symbols.[1] Figure 7–3 shows part of a mechanical plumbing plan.

STANDARD SYMBOLS AND FIGURES

Figure 7–4 shows the symbols used to show different types of water and waste lines, valves, and drains.

Symbol	Description
C.W.	Cold water
H.W.	Hot water
H.W.R.	Hot water return
W.L.	Waste line
V.L.	Vent line
S S	Sanitary sewer
C	Condensate line
S D	Storm drain
R.W.L.	Rain water leader
I.W.	Indirect waste
F	Fire line
G	Gas line
	Gate valve
	Globe valve
	Check valve
R	Relief line
	P&T relief valve
F.C.O.	Floor cleanout
F.D.	Floor drain
P.D.	Planter drain
R.D.	Roof drain
H.B.	Hose bibb
A.D.	Area drain

(Figure 7-4) Legend of plumbing symbols

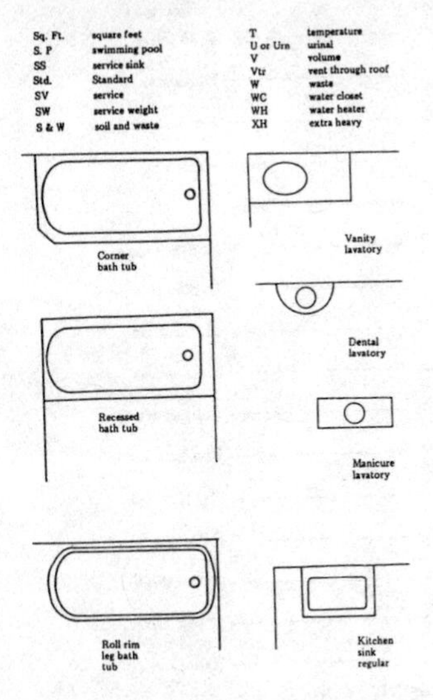

(Figure 7-5a) Symbols used for plumbing fixtures

Figure 7–5 shows the symbols for various plumbing fixtures.

Figure 7–6 gives the symbols used for fittings.

Note: Every plumber and apprentice must be thoroughly familiar with these symbols. Misinterpreting a blueprint or job order can lead to mistakes—mistakes that at worst may be dangerous, at best, time wasting and costly.

ISOMETRIC PLUMBING DRAWINGS

Isometric drawings are commonly used to show pipe and fitting sizes and placement. These drawings provide a three-dimensional picture of the plumbing system and are very useful to the plumber installing the system. (Plumbing inspectors often require an isometric drawing of a system or installation before issuing a permit.)

To draw an isometric sketch of a plumbing system, draw all vertical piping lines vertical in the sketch, but draw all horizontal lines at a 30° angle. (See Figure 7–7.)

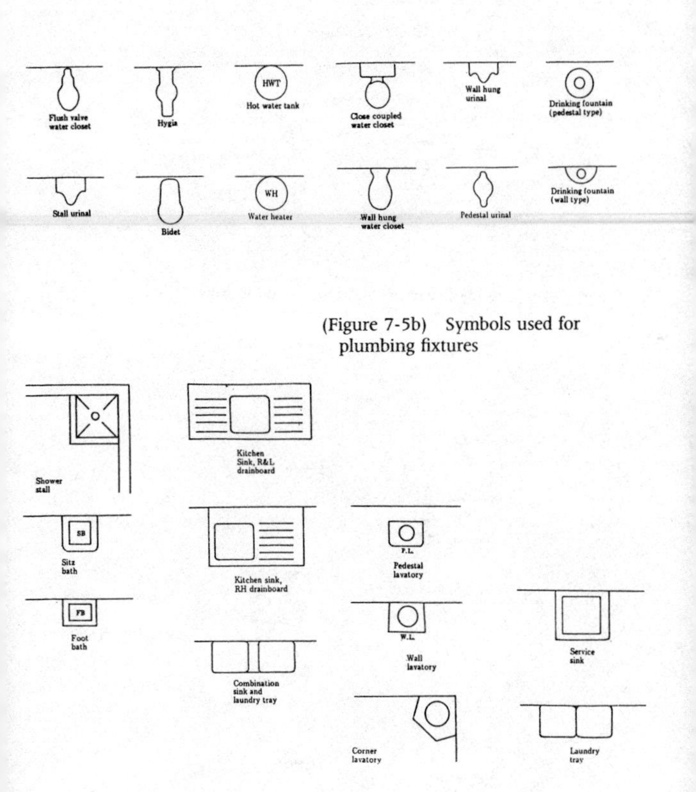

(Figure 7-5b) Symbols used for plumbing fixtures

(Figure 7-5c) Symbols used for plumbing fixtures

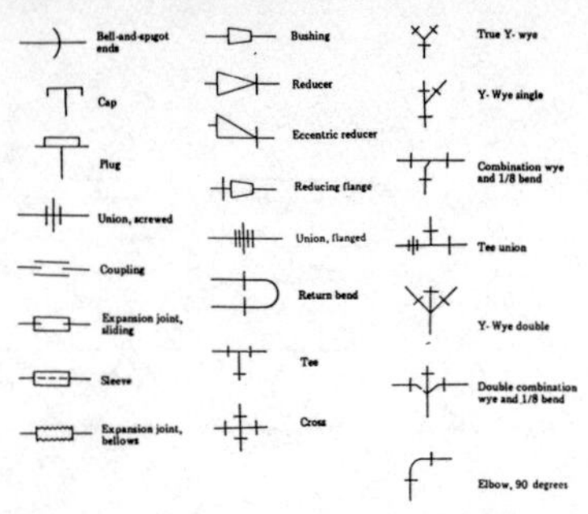

(Figure 7-6a) Symbols used for common pipe fittings

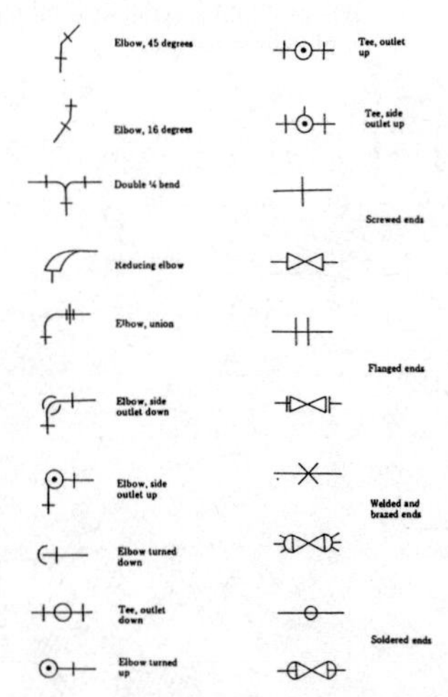

(Figure 7-6b) Symbols used for common pipe fittings

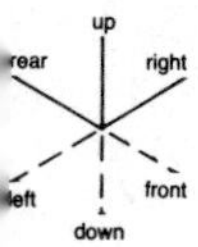

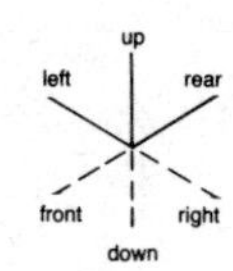

(Figure 7-7) Isometric sketches

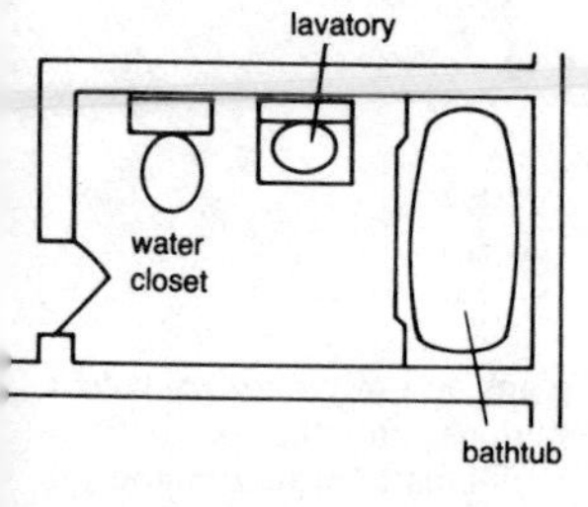

(Figure 7-8a) Plan, schematic drawing, and isometric drawing of a typical 3-fixture bathroom group

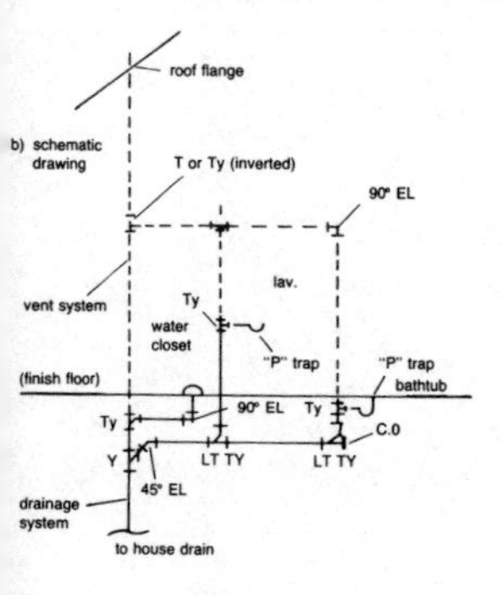

(Figure 7-8b) Plan, schematic drawing, and isometric drawing of a typical 3-fixture bathroom group

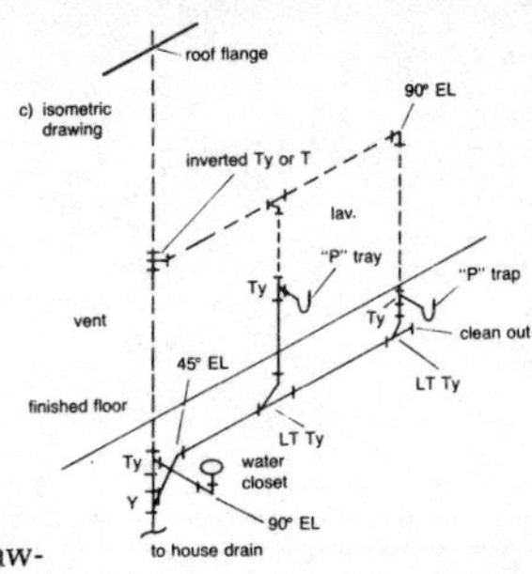

(Figure 7-8c) Plan, schematic drawing, and isometric drawing of a typical 3-fixture bathroom group

SUMMARY

Used together, the different types of plans and drawings provide a clear representation of a plumbing system or particular installation. Figure 7–8 shows the overall plan, a schematic drawing, and an isometric drawing of a typical three-fixture bathroom group.

Glossary

Abrasive any material that erodes another material by rubbing.

ABS (Acrylonitrile-Butadiene-Styrene) a plastic material used for drainage.

Adapter a fitting that joins pipes of different materials or different sizes.

Aerator device that adds air to water; it fills flowing water with bubbles to avoid splashing.

Air Chamber a device designed to absorb the shock of a fast closing valve or faucet.

Air Gap the unobstructed vertical distance through the air between the lowest outlet from any pipe or faucet supplying water to a tank, plumbing fixture, or other device and the flood level rim of the receptacle.

Anaerobic Bacteria bacteria that exist in the absence of free oxygen (air).

Anchor a special fastener used to attach pipes, fixtures, and other parts to the building structure.

Angle Valve a globe valve in which the inlet and outlet openings are at 90 degree angles to each other.

Apprentice Plumber a person who is learning the plumbing trade by practical experience.

Area Drain a drain installed to receive surface or rain water from an open area.

Asbestos Joint Runner a runner made of an asbestos rope and a clamp that holds molten lead in the bell of a cast pipe until it has cooled.

Back Fill material used to fill an excavated trench.

Backflow the flow of water in pipes in a reverse direction from that normally intended.

Backflow Connection any connection or arrangement whereby backflow may occur.

Backflow Preventer a device that prevents backflow into the potable water supply system.

Backing wood or other supports placed in the building walls to which plumbing fixtures and other equipment can be attached.

Back Vent a branch vent connected to the main vent stack and extending to a location near a fixture trap.

Backwater Valve a type of check valve installed to prevent the backflow of sewage from flooding the basement or lower levels of a building or dwelling.

Ballcock a valve or faucet controlled by a change in the water level; it is primarily associated with toilet tank operation.

Ball Valves a valve in which the flow of fluid is controlled by a rotating drilled ball that fits tightly against a resilient (flexible) seat in the valve body.

Basket Strainer a kitchen sink drain fitting, also called a *duo-strainer.*

Battery of Fixtures any group of two or more similar adjacent fixtures.

Bearing Partition an interior wall of a building that carries the load of the structure above in addition to its own weight.

Bell or Hub the enlarged end of some types of cast iron pipe that fits over the next pipe section.

Benchmark a fixed location of known elevation.

Bend a change of direction in piping.

Bib another name for faucet.

Bidet a bowl equipped with cold and hot running water used for bathing the external genitals and posterior parts of the body.

Blowoff the controlled discharge of excess pressure and temperature.

Blueprint drawings with accurate measurements that are used to install piping and building materials.

BOCA Building Officials Conference of America.

Bonnet the upper portion of the gate valve body.

Bracket Hanger a hanger supporting a wall-hung sink or fixture.

Branch an addition to the main pipe in a piping system.

Branch Vent a vent that connects a branch of the drainage piping to the main vent stack.

Braze a means of joining metal with an alloy having a melting point higher than common solder but lower than the metal being brazed.

Building Drain that part of the lowest piping of the drainage system that receives the discharge from soil, waste, and other draining pipes inside the walls of the building and conveys it to the building sewer.

Building Drainage System the complete system of pipes installed for the purpose of carrying waste water and sewage to septic or sanitary sewer systems.

Building Drain Branch a soil or waste pipe that extends horizontally from the building drain and receives only the discharge from fixtures on the same floor as the branch.

Building Main water supply piping that begins at the property line and ends in the building itself.

Building Sanitary Drain a building drain that conveys sewage only.

Building Sewer that part of the drainage system that extends from the end of the building drain and conveys its discharge to the public sewer system.

Building Storm Drain a building drain that conveys storm water only.

Building Storm Sewer a building sewer that conveys storm water, but not sewage.

Building Trap a trap placed in the building drain to prevent entry of sewer gases from the sewer main.

Burr a sharp, rough edge on a piece of pipe or tubing as a result of being cut.

Bushing a pipe fitting with both male and female threads; it is used to connect pipes of different sizes.

Cap a female pipe fitting that is closed at one end. It is used to close off the end of a piece of pipe or tubing.

Capillary Attraction the movement of liquid upward.

Cast-Iron Pipe any pipe made from cast iron.

Caulk the material used to seal joints.

Caulking a method of making a bell and spigot pipe joint watertight by packing it with oakum and lead.

Cesspool (Dry Well) a deep pit that receives liquid waste and permits the excess liquid to be absorbed into the ground.

Chain Wrench an adjustable tool for holding and turning large pipe up to 4 inches in diameter. A flexible chain replaces the steel jaws of 2 standard pipe wrenches.

Chalk Line a marking tool consisting of a string coated with chalk.

Change in Direction the term applied to the various turns that may be required in drainage pipes and other piping systems.

Chase (specifically, a Pipe Chase) a space or recess in the walls of a building where pipes are run.

Check Valve a device preventing backflow in pipes. Water can flow readily in one direction but any reversal of the flow causes the check valve to close.

Circuit Vent a branch vent that functions for two or more traps.

Cleanout a fitting with a removable plug that is placed in drainage pipe to allow entry into the system in order to relieve stoppages.

Close Nipple the shortest length of a given size pipe that is threaded on both ends.

Closet Bend an elbow drainage fitting connecting a water closet to the drainage system.

Closet Bolt a bolt used to attach a water closet securely to the closet flange.

Closet Spud the connector between the base of the ballcock assembly in a water closet tank and the water supply pipe.

Code a set of regulations that have been adopted by a governmental unit for the purpose of protecting the public health and safety. In plumbing, these codes regulate the quality of materials, the design and installation of plumbing systems, and the manner in which to test the systems.

Cold Chisel a hand tool used with a hammer to cut cast iron or concrete.

Common Vent a vent at the junction of two fixture drains, serving as a single vent for both fixtures.

Compression Faucet or Valve a faucet or valve designed to stop the flow of water by the action of a flat disc (washer) closing against a seat.

Continuous Vent a vertical vent that is a continuation of the drain to which it connects.

Continuous Waste the waste from two or three fixtures connected to a single trap.

Copper Pipe Straps straps made from copper that are used to secure copper pipe.

Corporation Stop a valve installed in the building water service line at the water main; it is also called corporation cock.

Counterflashing a flashing usually used on chimneys to prevent entry of moisture.

Coupling a pipe fitting containing female threads on both ends. Couplings are used to join two or more lengths of pipe in a straight run.

CPVC (Chlorinated Polyvinyl Choride) a type of plastic used to make pipe that will carry hot water, air, or chemicals.

Crawl Space the space between the floor framing and the ground in a building that has no basement.

Cross a pipe fitting with four female openings at right angles to one another.

Cross Connection any link between contaminated water and potable water in the supply system.

Crossover the connection of two piping runs in the same piping system or the connection of two different piping systems that contain potable water.

Crown of a Trap the point in a trap where the direction of flow changes from upward to downward.

Crown Weir the point in the curve of the trap directly below the crown.

Curb Box a cylindrical casting placed in the ground over the corporation stop. It extends to ground level and permits a special key to be inserted to turn off the corporation cock.

Curb Cock or Curb Stop a valve placed on the water service, usually near the curb line.

Dead End a branch of a drainage piping system that ends in a closed fitting.

Deep Seal Trap a trap located in the building drain to resist back pressure of sewer gas.

Developed Length the length of pipe and fittings measured along the center line.

Die a tool used to cut external threads by hand or machine.

Die Stock a tool used to turn dies when cutting external threads.

Dip of a Trap the lowest portion of the inside top surface of the trap.

Disposal Field see **Leach Bed.**

Dope a pipe joint compound.

Double Hub a cast-iron sewer pipe having a bell on both ends.

Downspout a vertical pipe made from sheet metal, copper, or plastic that carries water from the gutters to the ground or to a storm drain.

Drain any pipe that carries wastewater or waterborne wastes.

Drain, Building horizontal piping that connects the building drainage piping to the sanitary sewer or private sewage system.

Drainage Fitting a pipe fitting designed for use with drainage piping.

Drainage Fixture Unit (dfu) a measure of the probable discharge into the drainage system by various types of plumbing fixtures on the basis of one dfu being equal to 7.5 gallons per minute discharge. The drainage fixture-unit valve for a particular fixture depends on its volume rate of drainage discharge.

Drainage Piping all or any portion of the drainage piping system.

Drainage System the piping within public or private premises that conveys sewage or other liquid waste to a legal point of connection to a public sewer system or private disposal system.

Drains, Storm piping systems that carry water and rainwater from a building to the storm sewer.

Drinking Fountain a fixture that delivers a stream or jet of drinking water through a nozzle.

Drum Trap a trap whose main body is a cylinder with its axis vertical. This cylinder is larger in diameter than the inlet or outlet pipe.

Dry Vent any vent that does not carry waste water.

Dry Well a well filled with aggregate, designed to permit water to seep into the ground; used to receive rain water.

Ductility the property of a material that allows it to be formed into thin sections without breaking.

Duo Strainer see **Basket Strainer.**

Easement the right to use land owned by another for some specific purpose (for example, the right of a public utility or municipality to install service through a person's property).

Eccentric Fitting a pipe fitting in which the centerline of the openings is offset.

Effluent the outflow from sewage treatment equipment.

Eighth-Bend a pipe fitting that causes the run of pipe to make a 45 degree turn.

Elbow a pipe fitting having two openings which causes a run of pipe to change directions 90 degrees.

Erosion the gradual wearing away of material as a result of abrasive action.

Evaporation loss of water to the atmosphere.

Excavation Lines lines laid out on the job site to indicate where digging for foundation and piping is to be done.

Existing Work that part of the plumbing system that is in place when an addition or alteration is begun.

Expansion Joint a joint that permits pipe to move as a result of expansion.

Extra Heavy a term used to designate the heaviest and strongest grades of cast-iron and steel pipe.

Fall the amount of slope given to horizontal runs of pipe.

Faucet a valve whose purpose is to permit controlled amounts of water from the water pipe.

Female Thread any internal thread.

Ferrule a cast-iron fitting installed in the bell of a cast iron pipe.

Field Tile short lengths of clay pipe that are installed as subsurface drains.

Fill sand, gravel, or other loose earth.

Finishing the third major stage of the plumbing process.

Fittings the parts of the piping system that serve to join lengths of pipe.

Fixture a device such as a sink, lavatory, bathtub, water closet, or shower stall.

Fixture Branch the water supply piping that connects a fixture to the water supply piping.

Fixture Drain drainage piping including a trap that connects a fixture and a branch waste pipe.

Fixtures, Battery any two or more similar fixtures served by the same horizontal run of drainage piping.

Fixtures, Combination a fixture, such as a kitchen sink/laundry basin that is specifically designed to perform two or more functions.

Fixture Supply Pipe the water supply pipe that connects the fixture to the stub-out.

Fixture Unit a means of rating the amount of discharge from a given fixture so that the drainage piping is large enough to carry the required amount of waste. (Also: a flow of 1 cfm.)

Fixture Vent a part of the piping system that connects with the drainage piping near the point where the fixture trap is installed and extends to a point above the roof of the structure.

Flange a rim or collar attached to one end of a pipe to give support or a finished appearance.

Flange Nut a device that connects flared copper pipe to a threaded flare fitting.

Flashing materials such as copper or stainless steel which are installed as joints between roofs and walls and roofs and chimneys to prevent water from entering the structure.

Float Arm a thin rod threaded at each end that connects the float ball to the inlet valve of the ball cock assembly in a toilet tank.

Float Ball a metal or plastic ball used to control the inlet valve in water closet tanks.

Flooded a condition where liquid rises to the flood level of a fixture.

Flood Level the point in a fixture above which water overflows.

Floor Drain a fitting located in the floor to carry wastewater into the drainage piping.

Floor Flange a fitting attached at floor level to the end of a closet bend so that the water closet can be bolted to the drainage piping.

Flow Pressure the pressure in the water supply pipe near the faucet or water outlet while the faucet or water outlet is wide open and flowing.

Flow Rate the volume of water used by a plumbing fixture in a given amount of time. Usually expressed in gallons per minute (gpm).

Flush to clean by drenching with a large amount of water.

Flush Ball in a water closet tank assembly, the rubber ball-shaped object that controls the flow of water into the bowl.

Flush Bushing a pipe fitting used to reduce the diameter of a female-threaded pipe fitting.

Flushometer a valve that permits a preestablished amount of water to enter a fixture such as a water closet or urinal.

Flush Tank a receptacle designed to discharge, either manually or automatically, a predetermined quantity of water to fixtures for flushing purposes.

Flush Valve a device for flushing water closets and similar fixtures.

Flush Valve Seat the opening between the tank and bowl in a water closet against which the flush ball is fitted.

Flux a chemical substance that prevents oxides from forming on the surface of metals as they are heated for soldering, brazing, and welding.

Footing the part of the foundation of a building that rests directly on the ground. The footing distributes the weight of the building over a sufficiently large amount of ground so that the building will not settle excessively.

Force Cup a rubber cup attached to a wooden handle; it is used for unclogging water closets and drains. It is also called a plunger or "plumber's friend."

Foundation that part of a building which is below the first framed floor and includes the foundation wall and footing.

Foundation Drain piping around the base of the foundation to collect ground water and convey it into a sump.

Freezeless Water Faucet a water faucet designed to be installed through an exterior wall to prevent freezing.

Front Main Cleanout a plugged fitting located near the front wall of a building where the building drain leaves the building. The front main cleanout may be either inside or directly outside the building foundation wall.

Frostline the depth of frost penetration in the soil. This depth varies in different parts of the country depending on the normal temperature range.

Frost-Proof Closet a closet that has no water in the bowl and has the trap and flush valve installed below ground level, usually below the frost line.

Full-Bath a bathroom containing a water closet, a lavatory, and a bathtub.

Galvanized Iron iron that has been coated with zinc to prevent rust.

Garbage Disposal an electric grinding device used with water to grind food wastes into pulp and discharge the pulp into the drainage system.

Gasket any semihard material placed between two surfaces to make a watertight seal when the surfaces are drawn together by bolts or other fasteners.

Gate Valve a valve that utilizes a disc moving at a right angle to the flow of water to regulate the rate of flow. When a gate valve is fully opened, there is no obstruction to the flow of water.

Globe Valve a spherically shaped valve body that controls the flow of water with a compression disc. The disc, opened and closed by means of a stem, mates with a ground seat to stop water flow.

Grade or Pitch the fall (slope) of a line of pipe in reference to a horizontal plane. As applied to plumbing drainage, pitch is usually expressed as the fall in a fraction of an inch per foot length of pipe (for example, ¼ inch per foot).

Grease Interceptor a receptacle designed to separate and retain grease and fatty substances from wastes normally discharged from kitchens.

Ground Water water in the subsoil.

Hacksaw a metal-cutting saw with a replaceable blade.

Half-Bath a bathroom containing a water closet and a lavatory.

Handle Puller a tool for removing handles from faucets and valves.

Hanger a support for pipe.

Header a water supply pipe to which two or more branch pipes are connected to service fixtures.

Headroom space between the floor and the lowest pipe, duct, or part of the framing.

Horizontal Branch any horizontal pipe in the waste piping system that extends from a stack to the fixture trap.

Horizontal Pipe any pipe that is installed so that it makes an angle of less than 45 degrees from level.

Hose Bib a water faucet made with a threaded outlet for the attachment of a hose.

House Drain the horizontal part of the drainage piping that connects the piping system within the structure to the sanitary sewer or private sewage treatment equipment.

Hub the enlarged end of a hub and spigot cast-iron pipe.

Hydrant water supply outlet with a valve located below ground.

Increaser a fitting installed in a vent stack before the stack goes through the roof. It enlarges the stack or vent pipe.

Indirect Waste Pipe a waste pipe that does not connect directly with the drainage system, but discharges into it through a properly trapped fixture or receptacle.

Individual Vent a pipe, installed to vent a fixture trap, that connects with the vent system above the fixture it serves.

Industrial Waste a liquid waste resulting from the processes employed in industrial establishments.

Interceptor a receptacle designed and constructed to intercept, separate, and prevent the passage of detrimental floating or heavy solids.

Invert the lowest portion of the inside of any horizontal pipe.

Joint Runner a tool composed of asbestos rope and a clamp used in leading joints in horizontal runs of bell and spigot cast-iron pipe.

Journeyman Plumber a person who has acquired the requisite skill and knowledge necessary for the proper installation of plumbing. The requirements for this title are four years of training and experience under the supervision of a licensed master plumber, or the equivalent thereof in education, training and experience.

Keel a colored marking crayon used for marking pipe.

"K" Grade Copper Tube copper tubing suitable for installation underground.

Laundry Tray a fixed tub, installed in a laundry room of a home, that is supplied with cold and hot water and a drain connection and is used for washing clothes and other household items.

Lavatory a fixture designed for washing hands and face.

Lay Out the act of measuring and marking the location of something.

Layout the arrangement of a house, room, or part of a job.

Leach Bed a system of underground piping that permits absorption of liquid waste into the earth; also called disposal field or leach field.

Leader a pipe from a roof drain to a building storm drain.

Level a tool used to determine if something is horizontal or vertical.

"L" Grade Copper Tube a type of copper tube that may be used to supply potable water.

Line Level a small, lightweight level designed to be hung from a string line to determine if the line is horizontal.

Liquid Waste the liquid discharge from a plumbing fixture.

Long Quarter-Bend a 90-degree fitting with one section longer than the other.

Long-Sweep Fitting ,any drainage fitting that has a long radius curve at the bends.

Loop Vent a branch vent similar to a circuit vent except that it connects with the stack vent instead of the vent stack.

Lot Line the line(s) forming the legal boundary of a piece of property.

Main the principal pipe artery to which branches may be connected.

Main Sewer the large sewer to which the building drains of several houses are connected.

Main Vent the principle artery of the venting system to which vent branches may be connected.

Main Water Line the large water supply pipe to which branches are connected.

Male Thread threads on the outside of a pipe, fitting, or valve.

Malleable Iron cast iron which has been heat treated to reduce its brittleness.

Mallet a soft-face hammer (rawhide or plastic) used to drive parts without damaging them.

Manhole an opening in the sanitary or storm sewer system to permit access.

Masonry Bit a bit designed to drill holes in mortar, tile, and concrete.

Master Plumber a person who has had at least two years of experience as a journeyman plumber and is licensed as a master plumber and is engaged in the business of plumbing.

Meter Stop a valve used on a water main between the street and a water meter.

Miter Box a hardwood or metal saw guide. The sides are slotted to guide a hand saw for 45 and 90 degree cuts.

Mixing Faucet separate faucets having a common spout permitting control of the water temperature.

Moisture Barrier a material such as polyethylene that retards the passage of vapor or moisture into walls or through concrete floors.

Mop Basin a floor set service sink; also called a mop receptor.

Negative Pressure a pressure within a pipe that is less than atmospheric pressure.

Neoprene a synthetic rubber with superior resistance to oils; often used as gasket and washer material.

Nipples short lengths of pipe (usually less than 12 inches) with male threads on both ends that is used to join fittings.

No-Hub Pipe soil pipe that has smooth ends, but doesn't have a spigot or hub.

Nominal Size the approximate dimension(s) of standard material.

Nonbearing Wall a wall within a structure that supports no load other than its own weight.

Nonrising Stem Valve a gate valve in which the stem does not rise when the valve is opened.

Nozzle a fitting attached to the outlet of a pipe or hose that varies the volume of water and causes the shape of the stream of water to be changed to a spray of varying diameter.

Oakum loosely woven hemp rope that has been treated with oil or other waterproofing agent; it is used to caulk joints in a bell and spigot pipe and fittings.

Offset a combination of elbows or bends that permits a section of a pipe to be out of line but in a line parallel with its original alignment.

O-Ring a rubber seal used around stems of some valves to prevent water from leaking past.

Outside Wall any wall of a structure exposed to the weather on one side.

Overflow Tube a vertical tube in a water closet tank that prevents overfilling of the tank.

Oxidized Sewage sewage that has been exposed to oxygen to make the organic substances stable.

Packing a loosely packed waterproof material installed in the packing box of valves to prevent leaking around the stem.

Packing Nut a special nut holding the stem in a faucet or valve while compressing the packing.

Partition or Partition Wall an interior wall that divides spaces within a building.

Petcock a small ground key type valve used with soft copper tubing.

Pilot Light a relatively small flame that burns constantly and whose purpose is to ignite the main supply of gas.

Pipe a cylindrical conduit or conductor, the wall thickness of which is sufficient to receive a standard pipe thread.

Pipe Die a tool for cutting external pipe threads.

Pipe Joint Compound material used for sealing threaded pipe joints.

Pipe, Soil a pipe for conveying waste that contains fecal matter (human waste).

Pipe Strap a metal strap used to support or hold pipe in place.

Pipe, Vertical any pipe or part thereof that is installed in a vertical position.

Pipe, Waste a pipe that conveys only liquid and other waste, not fecal matter.

Pipe, Water Distribution pipes that carry water from the service pipe to fixtures in the building.

Pipe, Water Riser a water supply pipe that rises vertically from a horizontal pipe.

Pipe Wrench a wrench with adjustable, slightly curved, toothed jaws designed to grip pipe firmly as pressure is applied to the handle.

Pipes, Water Service that portion of the water piping that extends from the main to the meter.

Piping a generic term used to refer to all the pipes in a building.

Pitch the degree of slope or grade given a horizontal run of pipe.

Plug a pipe fitting with external threads and a head that is used for closing the opening in another fitting.

Plumb exactly perpendicular (vertical); at a right angle to the horizontal.

Plumb Bob a tool consisting of a weight suspended by a string. When allowed to hand freely, the string line will assume a position which is exactly vertical.

Plumber a person trained and experienced in the skill of plumbing.

Plumber's Friend a plunger, or force cup; a tool consisting of a rubber cup and handle used underwater to force a blockage through sewer lines.

Plumber's Furnace a heating source used to melt lead, heat soldering irons, or solder.

Plumbing the art of installing in buildings the pipes, fixtures, and other apparatus for bringing in the water supply and removing wastewater and water-carried waste.

Plumbing Appliance a special class of plumbing fixture intended to perform a special function.

Plumbing Fixture a receptacle for wastes that are ultimately discharged into the sanitary drain system.

Plumbing Inspector a person authorized to inspect plumbing and drainage for compliance with the code for the municipality.

Plumbing System the plumbing system of a building—including the water supply distributing pipes; the fixtures and fixture traps; the soil, waste, and vent pipes; the building drain and building sewer; and the storm water drainage—with its devices, appurtenances, and connections within and outside the building within the property line.

Plumbing Wall the wall in a building in which the plumbing pipes are installed.

Polyethylene a plastic used to make pipe and fittings for underground water systems.

Pool a permanently installed water receptacle used for swimming, diving, or bathing, designed to accomodate more than one person at a time.

Pop-Off Valve a safety valve that opens automatically when pressure and temperature exceeds a predetermined limit.

Porcelain a white ceramic material used for bathroom fixtures; it is also called vitreous enamel.

Port Control Faucet a single-handle, noncompression faucet that contains within the faucet body a port for both cold and hot water and some method of opening and closing these ports.

Positive Pressure a pressure within the sanitary drainage or vent piping system that is greater than atmospheric pressure.

Potable Water water from a public water supply approved by the State Department of Health or a private water supply that has been accepted by the Administrative Authority as satisfactory for human consumption.

Potable Water Supply System the water service pipe, the water distributing pipes, and the necessary connecting pipes, fittings, control valves, and all appurtenances within the building or outside the building within the property lines.

Precipitation the total measurable amount of water received in the form of snow, rain, hail, and sleet. It is usually expressed in inches per day, month, or year.

Pressure Head the amount of force or pressure created by a depth of one foot of water.

Pressure Regulator a valve that reduces water pressure in the supply piping.

Private Sewer a sewer system privately owned and not directly under the jurisdiction of a municipality or a public utility.

Propane hydrocarbon derived from crude petroleum and natural gas and used as a fuel for plumber's furnaces or torches.

P-Trap a trap commonly used on plumbing fixtures.

Public Potable Water Supply a water supply approved by the State Department of Health.

Public Sewer a sewer system approved by the State Department of Health and located in a street, alley, or other premises under the jurisdiction of a municipality or public utility.

Punch List a list, made by the home builder or owner near the end of construction, indicating what must be done before the house is completely finished and ready for occupancy.

Putty a soft, prepared mixture used to seal sink rims, water closet bases, and other places where a sealant is needed.

PVC (Polyvinyl Chloride) a type of plastic used to make plumbing pipe and fittings for water distribution, irrigation, and natural gas distribution.

Quarter-Bend a drainage pipe fitting that makes a 90-degree angle.

Rainwater Leader a pipe inside the building that conveys storm water from the roof to a storm drain; also called a conductor or downspout.

Reamer a tool used in reaming.

Reaming removing the burr from the inside of a pipe that has been cut.

Recovery Rate speed at which a water heater will heat cold water to the desired temperature.

Reducer a pipe fitting having one opening smaller than the other. Reducers are used to change from a relatively large diameter pipe to a smaller one.

Refill Tube a copper or rubber tube extending from the ballcock to the overflow tube in the water closet assembly.

Reinforcement Wire heavy woven wire placed in concrete to give added strength.

Reinforcing Rod embossed steel rods placed in concrete slabs, beams, or columns to increase their strength.

Relief Valve a safety device that automatically provides protection against excessive temperatures, excessive pressures, or both.

Relief Vent a branch from a vent stack, connected to a horizontal branch between the first fixture drain and a soil or waste stack.

Return Offset an offset that permits a pipe to be returned to its original alignment.

Rigid Copper Tubing hard copper used when installing water lines.

Rim the unobstructed open edge of the receptacle section of a plumbing fixture.

Riser the water supply pipe that extends vertically for the height of one full story or more and from which water is supplied to fixture branches.

Rising Stem a type of valve stem that moves up and down as the valve is opened and closed.

Roof Drain a drain installed in a flat or nearly flat roof to receive water and conduct it into a leader, downspout, or conductor.

Roof Jacket or Flange a jacket or flange installed on the roof terminals of vent stacks and stack vents to seal the opening and prevent rainwater from entering into the building around the vent pipe.

Rotating Ball Faucet a single-handed faucet that controls water flow and temperature with a channeled rotating plastic ball. Holes in the ball are aligned with orifices for hot and cold water.

Rough-In earliest stage of plumbing installation, sometimes divided into two stages: first rough brings water and sewer lines inside the building foundation; second rough is the installation of all piping that will be enclosed in the walls of the finished building.

Rough-In Measurements measurements that indicate where the water supply and waste piping must terminate in order to serve the fixtures which will be installed later.

Run one or more lengths of pipe that continue in a straight line.

Running Trap a stretch of pipe in which the inlet and outlet are at the same height and the waterway between them is lower than the bottom of either.

Saddle Fitting a fitting used to install a branch from an existing run of pipe.

Safety Valve a combination temperature and pressure relief valve generally installed in a hot water tank to prevent an explosion caused by overheating or excessive pressure inside the tank.

Sand Trap or Interception a device designed to allow sand and other heavy particles to settle out before water enters the water supply piping.

Sanitary Drainage Pipe pipes installed to remove waste-water and waterborne wastes.

Sanitary Sewage water and waterborne waste containing human excrement as well as other liquid household wastes.

Sanitary Sewer a sewer especially designed to carry sewage.

Sanitary T-Branch a drainage fitting having three openings and formed in the shape of a T.

Sanitary Y-Branch a drainage fitting shaped like a Y.

Scaffold any platform erected temporarily to support workers and materials while work is being done.

Scale Drawing a drawing of any object that has been carefully reduced to a fraction of the real size so that all parts are in the correct proportion.

Scuttle a small opening in a ceiling that provides access to an attic or roof.

Seal of a Trap the depth of water held in a trap under normal operating conditions.

Secondary Branch any branch off the primary branch of a building drain.

Self-Syphonage the loss of the seal of a trap as a result of removing the water from the trap. It is caused by the discharge of the fixture to which the trap is connected.

Septic Tank a watertight tank in a private waste disposal system that receives household sewage. Within the septic tank, solid matter is separated from the water before the water is discharged.

Service L (Street L) a 45-degree or 90-degree elbow with external threads on one end and internal threads on the other.

Service Pipe the water supply pipe from the main in the street or other source of supply to the building.

Service Sink a sink with a deep basin to accommodate a scrub pail. It is used for the filling and emptying of scrub pails, the rinsing of mops, and the disposal of cleaning water; it is also called a slop sink.

Sewage all water and waterborne waste discharged through the fixture.

Sewer a piping system designed to convey sewage.

Sewer, Building (House Sewer) horizontal sewage piping that extends from the building to the sewer main.

Sewer, Building Storm the piping from the building storm drain to the public storm sewer.

Sewer, Private a sewer owned and maintained privately. It may convey sewage from building(s) to a public sewer or to a privately owned sewage disposal system.

Sewer, Storm a sewer used to carry rainwater, surface water, or similar water wastes that do not include sanitary sewage.

Sewer Gas the mixture of vapors, odors, and gases found in sewers.

Shut-Off Valve a valve installed in a waterline whenever a cut off is required.

Side Outlet an opening at the side of a fitting; a T or Y fitting having one side opening.

Side Vent a vent connected to a drain at an angle of 45 degrees or less.

Sill Cock a faucet used on the outside of a building to which the garden hose can be attached.

Single Lever Faucet any of several types of washerless faucets using a single control and springs, balls, or cartridges to control the flow and temperature of the water.

Siphonage a partial vacuum created by the flow of liquids in pipes.

Size of Pipe the nominal dimension by which the pipe is designated; approximately equal to the inside diameter of the pipe.

Slab a large, flat, concrete section such as a basement floor, driveway, or patio.

Slip Coupling a pipe coupling that has no stop to prevent it from slipping over a pipe.

Slip Joint a connection in which one pipe slides inside another.

Slip Nut a nut used on P traps and similar connections. A gasket is compressed around the joint by the slip nut to form a watertight seal.

Slop Sink a deeper fixture than an ordinary sink, frequently installed in custodians' rooms.

Soil Pipe a pipe that conveys the discharge of water closets, or plumbing fixtures having similar functions, with or without discharges from other plumbing fixtures.

Soil, Stack the main vertical stack that receives and conveys the discharge from all plumbing fixtures.

Solder a metal alloy composed of tin and lead and used to join copper pipe and fittings.

Soldering Iron a tool composed of copper that is heated in a furnace and used to melt solder when joining pieces of metal.

Solder Joint the means of joining copper pipe to slip on fittings using solder.

Specifications a document that describes the quality of materials and the work quality required for a given building. Specifications are the plumber's source of information about the quality of the pipe, fixtures, etc., to be included in the plumbing system.

Spigot the plain end of a cast-iron pipe. The spigot is inserted into the bell end of the next pipe to make a watertight joint.

Splash Guard a specially formed block that is placed under the outlet of a downspout to prevent erosion of the soil.

Spout the end of a faucet that serves as a passageway for water out of the piping system.

Stack a general term for the vertical main of a system of soil, waste, or vent piping.

Stack Cleanout a plugged fitting located at the base of all soil or waste stacks.

Stack Vent the extension of a soil or waste stack above the highest connected horizontal branch.

Star Drill a tool made from steel that has a star-shaped chisel on one end and a face that is hit with a hammer on the other end. This tool is used to make holes in concrete and masonry block.

Stem the shaft of a faucet that holds the washer and to which the handle is attached.

Stock or Die Stock a tool used to turn a die when cutting threads.

Stop and Waste Valve a gate or compression-type valve that has a side opening, or port, and may be opened to allow water to drain from the piping supplied by the valve.

Stop Box or Curb Box an adjustable cast iron box that is brought up to grade with a removable iron cover.

Stopper a plug that controls wastewater drainage from a lavatory or bathtub, usually controlled remotely by a handle on the fixture; sometimes called a pop-up plug.

Storm Drain a drain that conveys rainwater, subsurface water, or other waste that does not need to be treated in a private or public sewage treatment facility.

Storm Sewer a sewer designed to convey only surface or storm water.

Storm Water the excess rainfall that runs off during or after a rain.

Storm Water Drainage System the piping system used for conveying rainwater or other precipitation to the storm sewer or other place of disposal.

S-Trap an S-shaped, water sealed trap sometimes used in plumbing. (Most water closet traps are S-traps.)

Strap Wrench a tool for gripping pipe. (The strap is made of nylon web treated with latex.)

Street L an elbow fitting with one male end and one female end; it is same as a service L.

Street T a T with one female and one male threaded opening, plus an outlet opening with female threads.

Subfloor a rough floor consisting of boards or plywood panels applied directly over the floor joist.

Subsoil Drain a drain that receives only subsurface water and conveys it to a storm drain.

Sump a tank or pit installed in the basement of a building to collect subsurface water so it can be pumped to a storm drain.

Sump Pump a rotary-type pump that lifts water from the sump into a drain pipe.

Supports, Hangers, Anchors devices for securing pipes to walls, ceilings, floors, or other structural members, and plumbing fixtures to floors or walls.

Survey a description of a piece of property including the measurements and marking of land.

Swage to increase or decrease the diameter of a pipe by using a special tool which is forced into or around the pipe.

Sweat Soldering a method of soldering in which the parts to be joined are first coated with a thin layer of solder and then joined while exposed to a flame.

Swing Joint a joint in a threaded pipe line that permits fittings to be installed in a close space.

Tamp to firmly compact earth during backfilling.

Tap a tool rotated by hand or machine to produce internal threads.

Tapered Reamer a tool for deburring and cleaning the inside ends of pipes.

Tapped T a cast-iron T with at least one branch tapped to receive a threaded pipe or fitting.

Temperature and Pressure Relief Valve a safety valve designed to protect against dangerous conditions by relieving high temperature and/or high pressure from a water heater.

Thermostat an automatic device consisting of a temperature sensing unit that turns an energy source on and off; it is used in heating and cooling.

Three-quarter Bath a bathtub containing a water closet, a lavatory, and a shower bath.

Three-Quarter S-Trap a trap shaped like three-fourths of the letter "S."

Trap a fitting or device designed and constructed to provide, when properly vented, a liquid seal that will prevent the passage of air without materially affecting the flow of liquid through it.

Trap Arm that portion of a plumbing fixture drain between the trap weir and the vent pipe connection.

Trap Seal the vertical distance between the crown weir and the dip of a trap that determines the depth of the water seal of a trap.

Trim the water supply and drainage fittings that are installed on the fixture to control the flow of water into the fixture and the flow of wastewater from the fixture to the sanitary drainage system.

Trunk Line the main piping from which building drains or water supply piping branch.

Tube a conduit or conductor of cylindrical shape, the wall thickness of which is less than that needed to receive a standard pipe thread.

Tubing any thin-walled pipe that can be bent easily.

Tubing Cutter a tool used to cut tubing.

Underground Piping piping in contact and covered with earth.

Union a fitting used to join two lengths of pipe which permits disconnecting of the two without cutting them.

Unit Vent one vent pipe that serves two or more traps.

Vacuum Breaker a device that prevents the formation of a vacuum in a water supply pipe; it is installed to prevent backflow.

Valve a fitting installed by plumbers on a piping system to control the flow of fluid within that system.

Vanity a bathroom fixture consisting of a lavatory set into or onto the top of a cupboard or cabinet.

Vapor Barrier a material that prevents moisture from penetrating a wall, ceiling, or floor.

Vent that part of the drain, waste, or vent piping that permits air to circulate and protects the seals in traps from siphonage and back pressure.

Vent, Circuit a vent installed where two similar fixtures discharge into a horizontal waste branch.

Vent, Common a vent that serves two or more fixture traps.

Venting, Individual the venting of each trap.

Vent, Looped a vent that drops below the flood rim of a fixture before being connected to the main vent.

Vent Pipe the pipe installed to ventilate a building drainage system and to prevent trap siphonage and back pressure.

Vent, Relief a vent installed at a point where the waste piping changes direction.

Vent Stack the vertical portion of the vent piping that extends through the roof of the building.

Vent System a pipe or pipes installed to provide a flow of air to or from a drainage system or to provide a circulation of air within such a system in order to protect trap seals from siphonage and back pressure.

Vent, Wet a pipe that serves as both a vent and a drain.

Vitreous Enamel see **Porcelain.**

Wall-Hung referring to a plumbing fixture supported by the wall.

Waste a liquid discharged from a fixture; the liquid contains no fecal matter.

Waste Pipe a pipe that conveys liquid waste that does not contain fecal matter.

Water Closet a toilet.

Water Conditioner a device used to remove dissolved minerals from water.

Water Cooler an electric appliance that combines a drinking fountain with a water cooling unit.

Water Distributing Pipe the piping that conveys water from a water service pipe to the fixture branch.

Water Distributing System the piping that conveys water from a service pipe to plumbing fixtures and other outlets.

Water Hammer the banging noise in pipes caused by a fast closing valve.

Water Heater an appliance for heating and supplying the hot water used within a building for purposes other than space heating.

Water Main a large water supply pipe, generally located near the street, that serves a large number of buildings.

Water Meter a device used to measure the amount of water in cubic feet or gallons that passes through the water service.

Water Outlets a discharge opening in a water supply system of a building or premises through which water can be obtained for the several purposes for which it is used by means of a faucet, valve, or other control mechanism.

Water Service Pipe the pipe conveying water from a water main or other source of water supply to the water distributing system of a building.

Water Softener an appliance that removes dissolved minerals (calcium and magnesium) from water by the process of ion exchange.

Water Supply Fixture Unit (WSFU) a common measure of the probable hydraulic demand on the water supply by various types of plumbing fixtures.

Water Supply System the water service pipe, water distributing system, fittings, and their accessories in or adjacent to any building, structure, or conveyance.

Wet Vent a soil or waste pipe serving as a vent.

Working Drawings drawings showing exactly how a building should be constructed.

Yarning Iron a tool used to pack oakum into bell-and-spigot pipe joints before they are leaded.

Yoke Vent a vent pipe installed from a soil or waste stack that connects to a vent stack at a higher elevation for the purpose of preventing pressure changes in the two stacks.

Y or Wye-Branch a section of pipe that joins the main run of pipe at an angle. The fitting that makes the joint is in the shape of the letter Y.

PART FOUR

APPENDIXES

1

Rules of Safety

As a general rule, safety is a matter of common sense. Each working location—whether it be a roof, ditch, crawl space, ladder or scaffolding—has its dangerous aspects, and the plumber must always exercise thought and care.

There are many federal and state laws pertaining to safety on the job. In 1971 the federal government enacted the National Occupational Safety and Health Act (OSHA), which made safety and health on the job a matter of law for all businesses and their employees. OSHA issues regulations, provides training programs for employers and employees, assists in establishing plans in compliance with legal regulations, and performs other functions to promote and implement safety and health standards on the job. OSHA and Department of Labor inspectors conduct on-site inspections to ensure that all safety laws are being complied with. If they find serious violations, they are empowered to shut down the job site.

In addition to federal and state agencies, many local agencies, trade organizations, and unions provide advice and assistance concerning the maintenance of safe working conditions and practices.

Specific tools, situations, and working locations present particular safety hazards. A few general safe working habits are listed before safety guidelines for particularly hazardous situations are discussed.

GENERAL SAFE WORKING HABITS

- Wear safety equipment.
- Observe all safety rules at the particular location.
- Be aware of any potential dangers in the specific situation.
- Keep tools in good condition.

SAFE DRESSING HABITS

- Do not wear clothing that can be ignited easily.
- Do not wear loose clothing, wide sleeves, ties, or jewelry (bracelets, necklaces) that can become caught in a tool or otherwise interfere with work. This caution is especially important when working with electrical machinery.
- Wear gloves to handle hot or cold pipes and fittings.
- Wear heavy-duty boots. Avoid wearing sneakers on the job. Nails can easily penetrate sneakers and can cause a serious injury (especially if the nail is rusty).
- Always tie shoelaces. Loose shoelaces can easily cause you to fall, possibly leading to injury to yourself or other workers.
- Wear a hard hat on major construction sites to protect the head from falling objects.

SAFE OPERATION OF GRINDERS

- Read the operating instructions before starting to use the grinder.
- Do not wear any loose clothing or jewelry.
- Wear safety glasses or goggles.
- Do not wear gloves while using the machine.
- Shut the machine off promptly when you are finished using it.

SAFE USE OF HAND TOOLS

- Use the right tool for the job.

- Read any instructions that come with the tool unless you are thoroughly familiar with its use.
- Wipe and clean all tools after each use. If any other cleaning is necessary, do it periodically.
- Keep tools in good condition. Chisels should be kept sharp and any mushroomed heads kept ground smooth; saw blades should be kept sharp; pipe wrenches should be kept free of debris and the teeth kept clean; etc.
- Do not carry small tools in your pocket, especially when working on a ladder or scaffolding. If you should fall, the tools may penetrate your body and cause serious injury.

SAFE USE OF ELECTRIC TOOLS

- Always use a three-prong plug with an electric tool.
- Read all instructions concerning the use of the tool (unless you are thoroughly familiar with its use).
- Make sure that all electrical equipment is properly grounded. Ground fault circuit interrupters (GFCI) are required by OSHA regulations in many situations.
- Use proper sized extension cords. (Undersized wires can burn out a motor, cause damage to the equipment, and present a hazardous situation.
- Never run an extension cord through water or through any area where it can be cut, kinked, or run over by machinery.
- Always hook up an extension cord to the equipment and then plug it into the main electrical outlet—not vice versa.
- Coil up and store extension cords in a dry area.

RULES FOR WORKING SAFELY IN DITCHES OR TRENCHES

- Be careful of underground utilities when digging.
- Do not allow people to stand on the top edge of a ditch while workers are in the ditch.
- Shore all trenches deeper than 4 feet.

- When digging a trench, be sure to throw the dirt away from the ditch walls (2 feet or more away).
- Be careful to see that no water gets into the trench. Be especially careful in areas with a high water table. Water in a trench can easily undermine the trench walls and lead to a cave in.
- Never work in a trench alone.
- Always have someone nearby—someone who can help you and locate additional help.
- Always keep a ladder nearby so you can exit the trench quickly if need be.
- Be watchful at all times. Be aware of any potentially dangerous situations. Remember, even heavy truck traffic nearby can cause a cave-in.

SAFETY ON ROLLING SCAFFOLDS

- Do not lay tools or other materials on the floor of the scaffold. They can easily move and you could trip over them, or they might fall, hitting someone on the ground.
- Do not move a scaffold while you are on it.
- Always lock the wheels when the scaffold is positioned and you are using it.
- Always keep the scaffold level to maintain a steady platform on which to work.
- Take no shortcuts. Be watchful at all times and be prepared for any emergencies.

WORKING SAFELY ON A LADDER

- Use a solid and level footing to set up the ladder.
- Use a ladder in good condition; do not use one that needs repair.
- Be sure step ladders are opened fully and locked.
- When using an extension ladder, place it at least ¼ of its length away from the base of the building.

- Tie an extension ladder to the building or other support to prevent it from falling or blowing down in high winds.
- Extend a ladder at least 3 feet over the roof line.
- Keep both hands free when climbing a ladder.
- Do not carry tools in your pocket when climbing a ladder. (If you fall, the tools could cut into you and cause serious injury.)
- Use the ladder the way it should be used. For example, do not allow two people on a ladder designed for use by one person.
- Keep the ladder and all its steps clean—free of grease, oil, mud, etc.—in order to avoid a fall and possible injury.

TO PREVENT FIRES (ESPECIALLY WHEN SOLDERING OR WELDING)

- Always keep fire extinguishers handy, and be sure that the extinguisher is full and that you know how to use it quickly.
- Be sure to disconnect and bleed all hoses and regulators used in welding, brazing, soldering, etc.
- Store cylinders of acetylene, propane, oxygen, and similar substances in an upright position in a well-vented area.
- Operate all air acetylene, welding, soldering, and related equipment according to the manufacturer's directions.
- Do not use propane torches or other similar equipment near material that can easily catch fire.
- Be careful at all times. Be prepared for the worst—and be ready to act.

2

Codes and Regulations

The installation, replacement, and repair of pipes and plumbing fixtures is governed by standards set by the National Standard Plumbing Code (NSPC). The regulations set forth in the code protect the public health and give guidelines for the handling of potable water and waste materials as well as provide information on pipes, fittings, installation practices, flow rates, and virtually all subjects and activities pertinent to plumbing.

The NSPC is reviewed and revised periodically to take into account the development of new materials (for example, plastics) for pipes and fittings, new tools and techniques, and any pertinent environmental and/or legal concerns.

Each state also has its own plumbing code, usually based on the NSPC. Many individual counties, municipalities, and towns also have their own codes, again usually based on state and federal guidelines.

Every plumber *must* be familiar with the National Standard Plumbing Code and refer to it frequently. Every plumber must also be familiar with the specific state and local requirements in the area in which he or she is working.

Any questions concerning the National Standard Plumbing Code should be addressed to

Code Secretary
National Standard Plumbing Code
P.O. Box 6808
Falls Church, VA 22046
(800) 533-7694 or (703) 237-8100

LICENSING EXAMINATIONS

Most states have regularly scheduled licensing examinations for persons wanting to become licensed master plumbers. (Requirements are different in each state, but, in general, to become a master plumber, you must have worked under the supervision of a master plumber for several years, first as an apprentice and then as a journeyman plumber. In some states licensing exams are also conducted by municipal and other local boards of health.

To find out the steps necessary for licensing in your own town, municipality, county, or state, contact your local town or city hall.

3
Useful Tables

The plumber must know the expected water demand at specific outlets, the minimum piping sizes required at specific fixtures, the types of valves to be used on different appliances and fixtures, and minimum trap and vent requirements as well as the basic characteristics of water in specific situations (for example, the boiling point and pressure), and the standard sizes of piping and tubing available. Much of this information is available to the plumber in easy-to-use tables. The following tables, although they do not contain all the data a plumber may need in a specific situation, contain much of the information a plumber may need in daily work.

The tables are given in alphabetical order according to the main word in the table title. (Many tables are listed under Water (e.g., boiling point, demand, and flow rate).

BUILDING DRAIN AND SEWER FIXTURE UNITS[1]

Diameter of Pipe	Maximum Number of Fixture Units That May be Connected to Any Portion of the Building Drain or the Building Sewer. Slope Per Foot			
	$\frac{1}{16}$-inch	$\frac{1}{8}$-inch	$\frac{1}{4}$-inch	$\frac{1}{2}$-inch
Inches				
2			21	26
$2\frac{1}{2}$			24	31
3			42[2]	50[2]
4		180	216	250
5		390	480	575
6		700	840	1,000
8	1,400	1,600	1,920	2,300
10	2,500	2,900	3,500	4,200
12	2,900	4,600	5,600	6,700
15	7,000	8,300	10,000	12,000

[1] On-site sewers that serve more than one building may be sized according to the current standards and specifications of the Administrative Authority for public sewers.
[2] Not over two water closets or two bathroom groups, except that in single family dwellings, not over three water closets or three bathroom groups may be installed.
COURTESY OF NATIONAL STANDARD PLUMBING CODE

VENT SIZE—TYPE OF FIXTURE

Type of Fixture	Minimum Size of Vent	dfu
Lavatory	$1\frac{1}{4}''$	1
Drinking fountain	$1\frac{1}{4}''$	1
Domestic sink	$1\frac{1}{4}''$	2
Shower stalls, domestic	$1\frac{1}{4}''$	2
Bathtub	$1\frac{1}{4}''$	2
Laundry tray	$1\frac{1}{4}''$	2
Service sink	$1\frac{1}{2}''$	3
Water closet	2″	6

COPPER TUBES—DIAMETERS

Nominal Size (in.)	Outside Dia. (in.) Types K, L, M, D.W.V.	Inside Diameter (in.)			
		Type K	Type L	Type M	Type D.W.V.
$\frac{1}{4}$	0.375	0.305	0.315		
$\frac{3}{8}$	0.500	0.402	0.430		
$\frac{1}{2}$	0.625	0.527	0.545		
$\frac{5}{8}$	0.750	0.652	0.666		
$\frac{3}{4}$	0.875	0.745	0.785		
1	1.125	0.995	1.025		
$1\frac{1}{4}$	1.375	1.245	1.265	1.291	1.295
$1\frac{1}{2}$	1.625	1.481	1.505	1.527	1.511
2	2.125	1.959	1.985	2.009	2.041
$2\frac{1}{2}$	2.625	2.435	2.465	2.495	
3	3.125	2.907	2.945	2.981	3.035
$3\frac{1}{2}$	3.625	3.385	3.425	3.459	
4	4.125	3.857	3.905	3.935	4.009
5	5.125	4.805	4.875	4.907	4.981
6	6.125	5.741	5.845	5.881	5.959
8	8.125	7.583	7.725	7.785	
10	10.125	9.449	9.625	9.701	
12	12.125	11.315	11.565	11.617	

SEPTIC TANK CAPACITY

Single family dwellings-number of bedrooms	Multiple dwelling units or apartments-one bedroom each	Other uses; maximum fixture units served	Minimum septic tank capacity in gallons
1–3		20	1000
4	2 units	25	1200
5 or 6	3	33	1500
7 or 8	4	45	2000
	5	55	2250
	6	60	2500
	7	70	2750
	8	80	3000
	9	90	3250
	10	100	3500

Extra bedroom, 150 gallons each.
Extra dwelling units over 10, 250 gallons each.
Extra fixture units over 100, 25 gallons per fixture unit.

Note: Septic tank sizes in this table include sludge storage capacity and the connection domestic food waste disposal units without further volume increase.

COURTESY OF NATIONAL STANDARD PLUMBING CODE

TAP AND DRILL SIZES
(American Standard Coarse)

Size of Drill	Size of Tap	Threads Per Inch	Size of Drill	Size of Tap	Threads Per Inch
7	$\frac{1}{4}$	20	$\frac{49}{64}$	$\frac{7}{8}$	9
F	$\frac{5}{16}$	18	$\frac{53}{64}$	$\frac{15}{16}$	9
$\frac{5}{16}$	$\frac{3}{8}$	16	$\frac{7}{8}$	1	8
U	$\frac{7}{16}$	14	$\frac{63}{64}$	$1\frac{1}{8}$	7
$\frac{27}{64}$	$\frac{1}{2}$	13	$1\frac{7}{64}$	$1\frac{1}{4}$	7
$\frac{31}{64}$	$\frac{9}{16}$	12	$1\frac{13}{64}$	$1\frac{3}{8}$	6
$\frac{17}{32}$	$\frac{5}{8}$	11	$1\frac{11}{32}$	$1\frac{1}{2}$	6
$\frac{19}{32}$	$\frac{11}{16}$	11	$1\frac{29}{64}$	$1\frac{5}{8}$	$5\frac{1}{2}$
$\frac{21}{32}$	$\frac{3}{4}$	10	$1\frac{9}{16}$	$1\frac{3}{4}$	5
$\frac{23}{32}$	$\frac{13}{16}$	10	$1\frac{11}{16}$	$1\frac{7}{8}$	5
			$1\frac{25}{32}$	2	$4\frac{1}{2}$

TRAP SIZES

Type of Fixture	Minimum Size of Trap in Inches
Clothes washer	$1\frac{1}{2}$
Bathtub with or without shower	$1\frac{1}{2}$
Bidet	$1\frac{1}{2}$
Dental unit or cuspidor	$1\frac{1}{4}$
Drinking fountain	$1\frac{1}{4}$
Dishwasher, domestic	$1\frac{1}{2}$
Dishwasher, commercial	2
Floor drain	2, 3, or 4
Lavatory	$1\frac{1}{4}$
Laundry tray (one or two compartment)	$1\frac{1}{2}$
Shower stall, domestic	$1\frac{1}{2}$
Sinks:	
Combination, sink and tray (with disposal unit)	$1\frac{1}{2}$
Combination, sink and tray (with one trap)	$1\frac{1}{2}$
Domestic with or without disposal unit	$1\frac{1}{2}$
Surgeon's	$1\frac{1}{2}$
Laboratory	$1\frac{1}{2}$
Flushrim or bedpan washer	3
Service	2 or 3
Pot or scullery	2
Soda fountain	$1\frac{1}{2}$
Commercial, flat rim, bar, or counter	$1\frac{1}{2}$
Wash, circular or multiple	$1\frac{1}{2}$
Urinals:	
Pedestal	3
Wall-hung	$1\frac{1}{2}$ or 2
Trough (per 6-foot section)	$1\frac{1}{2}$
Stall	2
Water closet	3

WATER SUPPLY FIXTURE UNITS AND FIXTURE BRANCH SIZES

Fixture[1]	Occupancy	Type of Supply Control	Load In Fixture Units[2]	Min. Size of Fixture Branch[4]
Bathroom group[3]	Private	Flushometer	8	—
Bathroom Group[3]	Private	Flush tank for closet	6	—
Bathtub	Private	Faucet	2	$\frac{1}{2}$
Bathtub	Public	Faucet	4	$\frac{1}{2}$
Clothes Washer	Private	Faucet	2	$\frac{1}{2}$
Clothes Washer	Public	Faucet	4	$\frac{1}{2}$
Combination Fixture	Private	Faucet	3	$\frac{1}{2}$
Kitchen Sink	Private	Faucet	2	$\frac{1}{2}$
Kitchen Sink	Public	Faucet	4	$\frac{1}{2}$
Laundry Trays (1 to 3)	Private	Faucet	3	$\frac{1}{2}$
Lavatory	Private	Faucet	1	$\frac{3}{8}$
Lavatory	Public	Faucet	2	$\frac{1}{2}$
Separate Shower	Private	Mixing Valve	2	$\frac{1}{2}$
Service Sink	Public	Faucet	3	$\frac{1}{2}$
Shower Head	Private	Mixing Valve	2	$\frac{1}{2}$
Shower Head	Public	Mixing Valve	4	$\frac{1}{2}$
Urinal-Pedestal	Public	Flushometer	10	1
Urinal	Public	Flushometer	5	$\frac{3}{4}$[5]
Urinal	Public	Flush Tank	3	$\frac{1}{2}$
Water Closet	Private	Flushometer	6	1
Water Closet	Private	Flush Tank	3	$\frac{1}{2}$
Water Closet	Public	Flushometer	10	1
Water Closet	Public	Flush Tank	5	$\frac{1}{2}$

Water supply outlets for items not listed above shall be computed at their maximum demand, but in no case less than the following values:

Fixture[4]	Number of Fixture Units	
	Private Use	Public Use
$\frac{3}{8}$	1	2
$\frac{1}{2}$	2	4
$\frac{3}{4}$	3	6
1	6	10

[1] For supply outlets likely to impose continuous demands, estimate continuous supply separately and add to total demand for fixtures.

[2] The given weights are for total demand. For fixtures with both hot and cold water supplies, the weights for maximum separate demands may be taken as $\frac{3}{4}$ the listed demand for the supply.

[3] A bathroom group for the purposes of this table consists of not more than one water closet, one lavatory, one bathtub, one shower stall or not more than one water closet, two lavatories, one bathtub or one separate shower stall.

[4] Nominal I.D. pipe size, inches

[5] Some may require larger sizes—see manufacturer instructions.

COURTESY NATIONAL STANDARD PLUMBING CODE

Size of soil or waste stack	Fixture Units Connected	Diameter of Vent Required (Inches)								
		1¼	1½	2	2½	3	4	5	6	8
		Maximum Length of Vent (Feet)								
Inches										
1½	8	50	150							
2	12	30	75	200						
2	20	26	50	150						
2½	42		30	100	300					
3	10		30	100	100	600				
3	30				60	200	500			
3	60			50	80	400				
4	100			35	100	260	1000			
4	200			30	90	250	900			
4	500			20	70	180	700			
5	200				35	80	350	1000		
5	500				30	70	300	900		
5	1100				20	50	200	700		
6	350				25	50	200	400	1300	
6	620				15	30	125	300	1100	
6	960					24	100	250	1000	
6	1900					20	70	200	700	
8	600						50	150	500	1300
8	1400						40	100	400	1200
8	2200						30	80	350	1100
8	3600						25	60	250	800
10	1000							75	125	1000
10	2500							50	100	500
10	3800							30	80	350
10	5600							25	60	250

COURTESY OF NATIONAL STANDARD PLUMBING CODE

STEAM PIPE EXPANSION
(Inches increase per 100 feet)

Temperature (Degrees F.)	Steel	Wrought Iron	Cast Iron	Brass and Copper
0	0	0	0	0
20	.15	.15	.10	.25
40	.30	.30	.25	.45
60	.45	.45	.40	.65
80	.60	.60	.55	.90
100	.75	.80	.70	1.15
120	.90	.95	.85	1.40
140	1.10	1.15	1.00	1.65
160	1.25	1.35	1.15	1.90
180	1.45	1.50	1.30	2.15
200	1.60	1.65	1.50	2.40
220	1.80	1.85	1.65	2.65
240	2.00	2.05	1.80	2.90
260	2.15	2.20	1.95	3.15
280	2.35	2.40	2.15	3.45
300	2.50	2.60	2.35	3.75
320	2.70	2.80	2.50	4.05
340	2.90	3.05	2.70	4.35
360	3.05	3.25	2.90	4.65
380	3.25	3.45	3.10	4.95
400	3.45	3.65	3.30	5.25
420	3.70	3.90	3.50	5.60
440	3.95	4.20	3.75	5.95
460	4.20	4.45	4.00	6.30
480	4.45	4.70	4.25	6.65
500	4.70	4.90	4.45	7.05
520	4.95	5.15	4.70	7.45
540	5.20	5.40	4.95	7.85
560	5.45	5.70	5.20	8.25
580	5.70	6.00	5.45	8.65
600	6.00	6.25	5.70	9.05
620	6.30	6.55	5.95	9.50
640	6.55	6.85	6.25	9.95
660	6.90	7.20	6.55	10.40
680	7.20	7.50	6.85	10.95
700	7.50	7.85	7.15	11.40
720	7.80	8.20	7.45	11.90
740	8.20	8.55	7.80	12.40
760	8.55	8.90	8.15	12.95
780	8.95	9.30	8.50	13.50
800	9.30	9.75	8.90	14.10

PIPE INFORMATION ON STANDARD WEIGHT

Nominal Diameter	Actual I.D.	Actual O.D.	Outside Circum. Inches	Outside Circum. Feet	Inside Area Sq. In.	Inside Area Sq. Ft.	Sq. Ft. O.D. Area Lin. Ft.	Gallons Lin. Ft.	Pounds Water Lin. Ft.	Wt. Pipe (Lb) Lin. Ft.
1/8	.269	.405	1.272	.106	.057	.0004	.106	.003	.024	.246
1/4	.364	.540	1.696	.141	.104	.0007	.141	.005	.045	.426
3/8	.493	.675	2.121	.177	.191	.0013	.177	.009	.082	.570
1/2	.622	.840	2.639	.220	.304	.0021	.220	.015	.131	.855
3/4	.824	1.050	3.299	.273	.533	.0037	.273	.027	.230	1.140
1	1.049	1.315	4.131	.343	.864	.006	.343	.044	.374	1.690
1 1/4	1.388	1.660	5.215	.433	1.496	.0103	.433	.077	.647	2.290
1 1/2	1.610	1.900	5.969	.497	2.036	.0141	.497	.105	.881	2.740
2	2.067	2.375	7.461	.622	3.356	.023	.622	.174	1.453	3.690
2 1/2	2.469	2.875	9.032	.751	4.778	.033	.751	.248	2.073	5.85
3	3.068	3.500	11.00	.843	7.393	.051	.843	.384	3.201	7.66
3 1/2	3.548	4.000	12.566	1.045	9.90	.068	1.045	.515	4.290	8.98
4	4.026	4.500	14.14	1.18	12.73	.088	1.178	.661	5.512	10.9
5	5.047	5.563	17.49	1.455	20.01	.139	1.455	1.039	8.662	14.9
6	6.065	6.625	20.81	1.73	28.89	.2	1.734	1.500	12.51	19.2
8	7.981	8.625	27.10	2.26	50.03	.35	2.258	2.598	21.66	28.9
10	10.020	10.750	33.772	2.81	78.85	.545	2.81	4.096	34.12	40.5
12	12.000	12.750	40.055	3.38	113.09	.984	3.38	5.88	48.96	49.56

PIPE (STANDARD WEIGHT)—DIAMETERS, THREADS

Nominal Size	Diameters (in inches)				Threads		Threads on Pipe	Thread-in
	O.D.	I.D.	Wood Auger for Pipe	Tap Drill	Number per Inch	Length on Pipe		
$\frac{1}{8}$	.405	.269	$\frac{1}{2}$	$\frac{21}{64}$	27	$\frac{3}{8}''$	10	$\frac{1}{4}''$
$\frac{1}{4}$	.540	.364	$\frac{5}{8}$	$\frac{29}{64}$	18	$\frac{9}{16}''$	10	$\frac{3}{8}''$
$\frac{3}{8}$	.675	.493	$\frac{11}{16}$	$\frac{19}{32}$	18	$\frac{9}{16}''$	10	$\frac{3}{8}''$
$\frac{1}{2}$	.840	.622	$\frac{15}{16}$	$\frac{23}{32}$	14	$\frac{3}{4}''$	$10\frac{1}{2}$	$\frac{1}{2}''$
$\frac{3}{4}$	1.050	.824	$1\frac{1}{8}$	$\frac{15}{16}$	14	$\frac{3}{4}''$	$10\frac{1}{2}$	$\frac{1}{2}''$
1	1.315	1.049	$1\frac{7}{16}$	$1\frac{3}{16}$	$11\frac{1}{2}$	1″	$11\frac{1}{2}$	$\frac{1}{2}''$
$1\frac{1}{4}$	1.660	1.380	$1\frac{3}{4}$	$1\frac{15}{16}$	$11\frac{1}{2}$	1″	$11\frac{1}{2}$	$\frac{1}{2}''$
$1\frac{1}{2}$	1.900	1.610	2	$1\frac{23}{32}$	$11\frac{1}{2}$	1″	$11\frac{1}{2}$	$\frac{1}{2}''$
2	2.375	2.067	$2\frac{1}{2}$	$2\frac{3}{16}$	$11\frac{1}{2}$	1″	$11\frac{1}{2}$	$\frac{1}{2}''$
$2\frac{1}{2}$	2.875	2.469	3	$2\frac{5}{8}$	8	$1\frac{1}{2}''$	12	$\frac{3}{4}''$
3	3.500	3.068	$3\frac{11}{16}$	$3\frac{1}{4}$	8	$1\frac{1}{2}''$	12	1″
$3\frac{1}{2}$	4.000	3.548	$4\frac{1}{4}$	$3\frac{3}{4}$	8	$1\frac{5}{8}''$	13	1″
4	4.500	4.025	$4\frac{11}{16}$	$4\frac{1}{4}$	8	$1\frac{5}{8}''$	13	1″
5	5.563	5.047	$5\frac{3}{4}$	$5\frac{5}{16}$	8	$1\frac{3}{4}''$	14	$1\frac{1}{4}''$
6	6.625	6.065	$6\frac{7}{8}$	$6\frac{5}{16}$	8	$1\frac{3}{4}''$	14	$1\frac{1}{4}''$
8	8.625	7.981	$8\frac{7}{8}$	$8\frac{3}{8}$	8	2″	16	$1\frac{1}{4}''$
10	10.75	10.02	11	$10\frac{5}{8}$	8	2″	16	$1\frac{1}{2}''$
12	12.75	12.00	13	$12\frac{5}{8}$	8	$2\frac{1}{2}''$	20	$1\frac{5}{8}''$

FIXTURE UNIT VALUES AND MINIMUM TRAP SIZE TABLE

Type of Fixture	Drainage Fixture Unit Value	Minimum Fixture Trap and Drain Size
Clothes washer (domestic use)	2	$1\frac{1}{2}$
Clothes washer (public use in groups of three or more	6 each	
Bathtub with or without shower	2	$1\frac{1}{2}$
Bidet	2	$1\frac{1}{2}$
Dental unit or cuspidor	1	$1\frac{1}{4}$
Drinking fountain	1	$1\frac{1}{4}$
Dishwasher, domestic	2	$1\frac{1}{2}$
Dishwasher, commercial	4	2
Floor drain with 2-inch waste	2	2
Floor drain with 3-inch waste	3	3
Floor drain with 4-inch waste	4	4
Lavatory	1	$1\frac{1}{4}$
Laundry tray (one or two compartment)	2	$1\frac{1}{2}$
Shower stall, domestic	2	$1\frac{1}{2}$
Shower (gang) per head	1	
SINKS:		
Combination, sink and tray (with disposal unit)	3	$1\frac{1}{2}$
Combination, sink and tray (with one trap)	2	$1\frac{1}{2}$
Domestic	2	$1\frac{1}{2}$
Domestic, with disposal unit	2	$1\frac{1}{2}$
Surgeons	3	$1\frac{1}{2}$
Laboratory	1	$1\frac{1}{2}$
Flushrim or bedpan washer	6	3
Service	3	2
Pot or scullery	4	2
Soda fountain	2	$1\frac{1}{2}$
Commercial, flat rim, bar or counter	3	$1\frac{1}{2}$
Wash, circular or multiple (per set of faucets)	2	$1\frac{1}{2}$
URINAL Pedestal or Wall-hung, with 3-inch trap (blowout and syphon jet)	6	3
Wall-hung with 2-inch trap	4	2
Wall-hung with $1\frac{1}{2}$-inch trap	2	$1\frac{1}{2}$
Trough (per 6-foot section)	2	$1\frac{1}{2}$
Stall	3	2
WATER CLOSET	6	3
Unlisted Fixture or Trap Size		
$1\frac{1}{4}$ inch	1	
$1\frac{1}{2}$ inch	2	
2 inch	3	
$2\frac{1}{2}$ inch	4	
3 inch	5	
4 inch	6	

COURTESY OF MINNESOTA PLUMBING CODE

FIXTURE LOAD VALUES

Fixture	Occupancy	Type of Supply Control	Load Values, in Water Supply Fixture Units		
			Cold	Hot	Total
Water closet	Public	Flush valve	10.		10.
Water closet	Public	Flush tank	5.		5.
Urinal	Public	1″ flush valve	10.		10.
Urinal	Public	¾″ flush valve	5.		5.
Urinal	Public	Flush tank	3.		3.
Lavatory	Public	Faucet	1.5	1.5	2.
Bathtub	Public	Faucet	3.	3.	4.
Shower head	Public	Mixing valve	3.	3.	4.
Service sink	Offices, etc.	Faucet	2.25	2.25	3.
Kitchen sink	Hotel, restaurant	Faucet	3.	3.	4.
Drinking fountain	Offices, etc.	⅜″ valve	0.25		0.25
Water closet	Private	Flush valve	6.		6.
Water closet	Private	Flush tank	3.		3.0
Lavatory	Private	Faucet	0.75	0.75	1.
Bathtub	Private	Faucet	1.5	1.5	2.
Shower stall	Private	Mixing valve	1.5	1.5	2.
Kitchen sink	Private	Faucet	1.5	1.5	2.
Laundry trays (1 to 3)	Private	Faucet	2.25	2.25	3.
Combination fixture	Private	Faucet	2.25	2.25	3.
Dishwashing machine	Private	Automatic		1.	1.
Laundry machine (8 lb)	Private	Automatic	1.5	1.5	2.
Laundry machine (8 lb)	Public or General	Automatic	2.25	2.25	3.
Laundry machine (16 lb)	Public or General	Automatic	3.	3.	4.

Note: For fixtures not listed, loads should be assumed by comparing the fixture to one listed using water in similar quantities and at similar rates. The assigned loads for fixtures with both hot and cold water supplies are given for separate hot and cold water loads and for total load, the separate hot

DRAINAGE FIXTURE UNIT VALUES FOR VARIOUS PLUMBING FIXTURES

Type of Fixture or Group of Fixtures	Drainage Fixture Unit Value
Automatic clothes washer (2″ standpipe)	3
Bathroom group consisting of a water closet, lavatory and bathtub or shower stall:	6
Bathtub[1] (with or without overhead shower)	2
Bidet	1
Clinic Sink	6
Combination sink-and-tray with food waste grinder	4
Combination sink-and-tray with one 1½″ trap	2
Combination sink-and-tray with separate 1½″ traps	3
Dental unit or cuspidor	1
Dental lavatory	1
Drinking fountain	½
Dishwasher, domestic	2
Floor drains with 2″ waste	3
Kitchen sink, domestic, with one 1½″ trap	2
Kitchen sink, domestic, with food waste grinder	2
Kitchen sink, domestic, with food waste grinder and dishwasher 1½″ trap	3
Kitchen sink, domestic, with dishwasher 1½″ trap	3
Lavatory with 1¼″ waste	1
Laundry tray (1 or 2 compartments)	2
Shower stall, domestic	2
Showers (group) per head[2]	2
Sinks:	
Surgeon's	3
Flushing rim (with valve)	6
Service (trap standard)	3
Service (P trap)	2
Pot, scullery, etc.[2]	4
Urinal, pedestal, syphon jet blowout	6
Urinal, wall lip	4
Urinal, stall, washout	4
Wash sink (circular or multiple) each set of faucets	2
Water closet, private	4
Water closet, public	6
Fixtures not listed above:	
Trap Size 1¼″ or less	1
Trap Size 1½″	2
Trap Size 2″	3
Trap Size 2½″	4
Trap Size 3″	5
Trap Size 4″	6

[1] A shower head over a bathtub does not increase the fixture unit value.
[2] See the table on Fixture Load Values.
COURTESY OF NATIONAL STANDARD PLUMBING CODE

MEASUREMENT CONVERSION FACTORS

To Change	To	Multiply by
Inches	Feet	0.0833
Inches	Millimeters	25.4
Feet	Inches	12
Feet	Yards	0.3333
Yards	Feet	3
Square inches	Square feet	0.00694
Square feet	Square inches	144
Square feet	Square yards	0.11111
Square yards	Square feet	9
Cubic inches	Cubic feet	0.00058
Cubic feet	Cubic inches	1728
Cubic feet	Cubic yards	0.03703
Cubic yards	Cubic feet	27
Cubic inches	Gallons	0.00433
Cubic feet	Gallons	7.48
Gallons	Cubic inches	231
Gallons	Cubic feet	0.1337
Gallons	Pounds of water	8.33
Pounds of water	Gallons	0.12004
Ounces	Pounds	0.0625
Pounds	Ounces	16
Inches of water	Pounds per square inch	0.0361
Inches of water	Inches of mercury	0.0735
Inches of water	Ounces per square inch	0.578
Inches of water	Pounds per square foot	5.2
Inches of mercury	Inches of water	13.6
Inches of mercury	Feet of water	1.1333
Inches of mercury	Pounds per square inch	0.4914
Ounces per square inch	Inches of mercury	0.127
Ounces per square inch	Inches of water	1.733
Pounds per square inch	Inches of water	27.72
Pounds per square inch	Feet of water	2.310
Pounds per square inch	Inches of mercury	2.04
Pounds per square inch	Atmospheres	0.0681
Feet of water	Pounds pe square inch	0.434
Feet of water	Pounds per square foot	62.5
Feet of water	Inches of mercury	0.8824
Atmospheres	Pounds per square inch	14.696
Atmospheres	Inches of mercury	29.92
Atmospheres	Feet of water	34
Long tons	Pounds	2240
Short tons	Pounds	2000
Short tons	Long tons	0.89285

VOLUME MEASURE EQUIVALENTS

1 gallon (gal.) = 0.133681 cubic foot (cu. ft.)
1 gallon (gal.) = 231 cubic inches (cu. in.)

WATER WEIGHT

1 cu. ft. at 50°F. weighs 62.41 lb.
1 gal. at 50°F. weighs 8.34 lb.
1 cu. ft. of ice weighs 57.2 lb.
Water is at its greatest density at 39.2°F.
1 cu. ft. at 39.2°F. weighs 62.43 lb.

WATER—BOILING POINT AT VARIOUS PRESSURES

Vacuum, In Inches of Mercury	Boiling Point	Vacuum, In Inches of Mercury	Boiling Point
29	76.62	7	198.87
28	99.93	6	200.96
27	114.22	5	202.25
26	124.77	4	204.85
25	133.22	3	206.70
24	140.31	2	208.50
23	146.45	1	210.25
22	151.87	Gauge Lb.	
21	156.75	0	212.
20	161.19	1	215.6
19	165.24	2	218.5
18	169.00	4	224.4
17	172.51	6	229.8
16	175.80	8	234.8
15	178.91	10	239.4
14	181.82	15	249.8
13	184.61	25	266.8
12	187.21	50	297.7
11	189.75	75	320.1
10	192.19	100	337.9
9	194.50	125	352.9
8	196.73	200	387.9

WATER PRESSURE TO FEET HEAD

Pounds Per Square Inch	Feet Head	Pounds Per Square Inch	Feet Head
1	2.31	100	230.90
2	4.62	110	253.98
3	6.93	120	277.07
4	9.24	130	300.16
5	11.54	140	323.25
6	13.85	150	346.34
7	16.16	160	369.43
8	18.47	170	392.52
9	20.78	180	415.61
10	23.09	200	461.78
15	34.63	250	577.24
20	46.18	300	692.69
25	57.72	350	808.13
30	69.27	400	922.58
40	92.36	500	1154.48
50	115.45	600	1385.39
60	138.54	700	1616.30
70	161.63	800	1847.20
80	184.72	900	2078.10
90	207.81	1000	2309.00

Note: One pound of pressure per square inch of water equals 2.309 feet of water at 62° Fahrenheit. Therefore, to find the feet head of water for any pressure not given in the table above, multiply the pressure pounds per square inch by 2.309.

TEMPERATURE CONVERSIONS FROM DEGREES FAHRENHEIT TO APPROXIMATE DEGREES CELSIUS*

Fahrenheit (°F)	Celsius (°C)
32°	0°
40°	4°
50°	10°
60°	15°
70°	21°
80°	26°
90°	32°
100°	38°
120°	49°
140°	60°
160°	71°
180°	82°
200°	93°
212°	100°

***The formulas for converting:**

$$C = \frac{5}{9}(F - 32)$$
$$F = \frac{9}{5}(C) + 32$$

WATER FEET HEAD TO PSI

Feet Head	Pounds Per Square Inch	Feet Head	Pounds Per Square Inch
1	.43	100	43.31
2	.87	110	47.64
3	1.30	120	51.97
4	1.73	130	56.30
5	2.17	140	60.63
6	2.60	150	64.96
7	3.03	160	69.29
8	3.46	170	73.63
9	3.90	180	77.96
10	4.33	200	86.62
15	6.50	250	108.27
20	8.66	300	129.93
25	10.83	350	151.58
30	12.99	400	173.24
40	17.32	500	216.55
50	21.65	600	259.85
60	25.99	700	303.16
70	30.32	800	346.47
80	34.65	900	389.78
90	38.98	1000	433.00

Note: One foot of water at 62° Fahrenheit equals .433 pound pressure per square inch. To find the pressure per square inch for any feet head not given in the table above, multiply the feet head by .433.

MASS CONVERSIONS FROM OUNCES AND POUNDS TO KILOGRAMS

Ounces	Kilograms	Pounds	Kilograms
1	.028	1	.454
2	.057	2	.907
3	.085	3	1.361
4	.113	4	1.814
5	.142	5	2.268
6	.170	6	2.722
7	.198	7	3.175
8	.227	8	3.629
9	.255	9	4.082
10	.283	10	4.536
11	.312	25	11.34
12	.340	50	22.68
13	.369	75	34.02
14	.397	100	45.36
15	.425		
16 (1 lb.)	.454		

COURTESY OF NATIONAL STANDARD PLUMBING CODE

SQUARE MEASURE CONVERSIONS FROM SQUARE INCHES TO APPROXIMATE SQUARE CENTIMETERS AND SQUARE FEET TO APPROXIMATE SQUARE METERS

Square Inches	Square Centimeters	Square Feet	Square Meters
1	6.5	1	.0925
2	13.0	2	.1850
3	19.5	3	.2775
4	26.0	4	.3700
5	32.5	5	.4650
6	39.0	6	.5550
7	45.5	7	.6475
8	52.0	8	.7400
9	58.5	9	.8325
10	65.0	10	.9250
25	162.5	25	2.315
50	325.0	50	4.65
100	650.0	100	9.25

FLOW RATE EQUIVALENTS

1 gpm[1]	= 0.134 cu. ft. per min.
	= 500 lb. per hr. × sp. gr[2]
500 lb. per hr.	= 1 gpm ÷ sp. gr.
1 cu. ft. per min. (cfm)	= 448.8 gal. per hr. (gph)[3]

[1] gpm = gallons per minute
[2] sp. gr. = specific gravity
[3] gph = gallon per hour

FLOW-RATE CONVERSIONS—FROM GALLONS PER MINUTE (GPM) TO APPROXIMATE LITERS PER MINUTE

GPM	Liters/Minute
1	3.75
2	6.50
3	11.25
4	15.00
5	18.75
6	22.50
7	26.25
8	30.00
9	33.75
10	37.50

COURTESY NATIONAL STANDARD PLUMBING CODE

PRESSURE OR STRESS CONVERSIONS FROM POUNDS PER SQUARE INCH AND POUNDS PER SQUARE FOOT TO KILOPASCALS

PSI	Kilopascals	PSF	Kilopascals
1	6.895	1	.0479
2	13.790	2	.0958
3	20.685	3	.1437
4	27.580	4	.1916
5	34.475	5	.2395
6	41.370	6	.2874
7	48.265	7	.3353
8	55.160	8	.3832
9	62.055	9	.4311
10	68.950	10	.4788
25	172.375	25	1.1971
50	344.75	50	2.394
75	517.125	75	3.5911
100	689.50	100	4.788

COURTESY OF NATIONAL STANDARD PLUMBING CODE

WATER PRESSURE RANGE—46 TO 60 PSI

Meter and Water Service (in inches)	Building Supply and Branches (in inches)	Maximum Allowable Length in Feet		
		150	200	250
3/4	1/2	5	4	3
3/4	3/4	14	11	9
3/4	1	28	23	21
1	1	30	25	23
1	1 1/4	52	44	39
1 1/2	1 1/4	66	52	44
1 1/2	1 1/2	128	105	90
2	1 1/2	150	117	98
1 1/2	2	272	240	220
2	2	368	318	280
2	2 1/2	535	500	470

LENGTH CONVERSION FROM INCHES AND FEET TO METERS AND MILLIMETERS

Inches	Meters (m)	Millimeters (mm)	Feet	Meters (m)	Millimeters (mm)
1/8	0.003	3.17	1	0.305	304.8
1/4	0.006	6.35	2	0.610	609.6
3/8	0.010	9.52	3 (1 yd.)	0.914	914.4
1/2	0.013	12.6	4	1.219	1 219.2
5/8	0.016	15.87	5	1.524	1 524.0
3/4	0.019	19.05	6 (2 yd.)	1.829	1 828.8
7/8	0.022	22.22	7	2.134	2 133.6
1	0.025	25.39	8	2.438	2 438.2
2	0.051	50.79	9 (3 yd.)	2.743	2 743.2
3	0.076	76.20	10	3.048	3 048.0
4	0.102	101.6	20	6.096	6 096.0
5	0.127	126.9	30 (10 yd.)	9.144	9 144.0
6	0.152	152.4	40	12.19	12 192.0
7	0.178	177.8	50	15.24	15 240.0
8	0.203	203.1	60 (20 yd.)	18.29	18 288.0
9	0.229	228.6	70	21.34	21 336.0
10	0.254	253.9	80	24.38	24 384.0
11	0.279	279.3	90 (30 yd.)	27.43	27 432.0
12	0.305	304.8	100	30.48	30 480.0

COURTESY NATIONAL STANDARD PLUMBING CODE

LENGTH/TIME CONVERSIONS FROM FEET PER SECOND TO APPROXIMATE METERS PER SECOND

Feet Per Second	Meters Per Second
1	.3050
2	.610
3	.915
4	1.220
5	1.525
6	1.830
7	2.135
8	2,440
9	2,754
10	3.05

COURTESY OF NATIONAL STANDARD PLUMBING CODE

WATER FLOW—MINIMUM AND MAXIMUM RATES PER OUTLET

Fixture	Flow Rate Minimum	GPM Maximum
Lavatory	2	4
Sink	4	8
Bathbtub	6	—
Laundry Tray	5	—
Shower	4	8
Water Closets		
Tank Type	3	6
Blowout Action / Jet Action	Depends on Flow Pressure	
Drinking Fountain	0.75	2.0
Wall Hydrant	5	—

COURTESY NATIONAL STANDARD PLUMBING CODE

WATER FLOW RATE

Fixture	Flow Rate (gpm)
Ordinary basin faucet	2.0
Self-closing basin faucet	2.5
Sink faucet, ⅜″	4.5
Sink faucet, ½″	4.5
Bathtub faucet	6.0
Laundry tub cock, ½″	5.0
Shower	5.0
Ball cock for water closet	3.0
Flushometer valve for water closet	15–35
Flushometer valve for urinal	15.0
Drinking fountain	0.75
Sill cock or wall hydrant	5.0

SEEPAGE PITS—EFFECTIVE ABSORPTION AREA FOR EACH 100 GALLONS OF SEWAGE PER DAY

Time in Minutes for 1-inch Drop	Effective Absorption Area Square Feet
1	32
2	40
3	45
5	56
10	75
15	96
20	108
25	139
30	167

COURTESY OF NATIONAL STANDARD PLUMBING CODE

FIXTURE TRAP MAXIMUM DISTANCES

Size of Fixture Drain in Inches	Distance—Trap to Vent
1¼	2 ft. 6 in.
1½	3 ft. 6 in.
2	5 ft.
3	6 ft.
4	10 ft.

COURTESY OF MINNESOTA PLUMBING CODE

Note: The developed length between the trap of the water closet or similar fixture and its vent shall not exceed four (4) feet.

TRAP ARM—MAXIMUM LENGTH

Diameter of Trap Arm Inches	Length—Trap to Vent
1¼	3′-6″
1½	5′
2	8′
3	10′
4	12′

COURTESY NATIONAL STANDARD PLUMBING CODE

FIXTURE BRANCH PIPING—MINIMUM SIZE

Fixture or Device	Size, (in.)
Bathtub	$\frac{1}{2}$
Combination sink and laundry tray	$\frac{1}{2}$
Drinking fountain	$\frac{3}{8}$
Dishwashing machine (domestic)	$\frac{1}{2}$
Kitchen sink (domestic)	$\frac{1}{2}$
Kitchen sink (commercial)	$\frac{3}{4}$
Lavatory	$\frac{3}{8}$
Laundry tray (1, 2, or 3 compartments)	$\frac{1}{2}$
Shower (single head)	$\frac{1}{2}$
Sink (service, slop)	$\frac{1}{2}$
Sink (flushing rim)	$\frac{3}{4}$
Urinal (1″ flush valve)	1
Urinal ($\frac{3}{4}$″ flush valve)	$\frac{3}{4}$
Urinal (flush tank)	$\frac{1}{2}$
Water closet (flush tank)	$\frac{3}{8}$[1]
Water closet (flush valve)	1
Hose bib	$\frac{1}{2}$
Wall hydrant or sill cock	$\frac{1}{2}$

[1] Fixtures may require larger sizes—see manufacturer's instructions.
COURTESY OF NATIONAL STANDARD PLUMBING CODE

WATER—DEMAND AT INDIVIDUAL OUTLETS

Type of Outlet	Demand, (gpm)
Ordinary lavatory faucet	2.0
Self-closing lavatory faucet	2.5
Sink faucet, $\frac{3}{8}$″ or $\frac{1}{2}$″	4.5
Sink faucet, $\frac{3}{4}$″	6.0
Bath faucet, $\frac{1}{2}$″	5.0
Shower head, $\frac{1}{2}$″	5.0
Laundry faucet, $\frac{1}{2}$″	5.0
Ballcock in water closet flush tank	3.0
1″ flush valve (25 psi flow pressure)	35.0
1″ flush valve (15 psi flow pressure)	27.0
$\frac{3}{4}$″ flush valve (15 psi flow pressure)	15.0
Drinking fountain jet	0.75
Dishwashing machine (domestic)	4.0
Laundry machine (8 or 16 lb.)	4.0
Aspirator (operating room or laboratory)	2.5
Hose bib or sill cock ($\frac{1}{2}$″)	5.0

COURTESY OF NATIONAL STANDARD PLUMBING CODE

WATER DEMAND ESTIMATES

Supply Systems Predominantly for Flush Tanks		Supply Systems Predominantly for Flushometers	
Load	Demand	Load	Demand
(Water Supply Fixture Units)	(Gallons per Minute)	(Water Supply Fixture Units)	(Gallons per Minute)
6	5		
8	6.5		
10	8	10	27
12	9.2	12	28.6
14	10.4	14	30.2
16	11.6	16	31.8
18	12.8	18	33.4
20	14	20	35
25	17	25	38
30	20	30	41
35	22.5	35	43.8
40	24.8	40	46.5
45	27	45	49
50	29	50	51.5
60	32	60	55
70	35	70	58.5
80	38	80	62
90	41	90	64.8
100	43.5	100	67.5
120	48	120	72.5
140	52.5	140	77.5
160	57	160	82.5
180	61	180	87
200	65	200	91.5
225	70	225	97
250	75	250	101
275	80	275	105.5
300	85	300	110
400	105	400	126
500	125	500	142
750	170	750	178
1,000	208	1,000	208
1,250	240	1,250	240
1,500	267	1,500	267
1,750	294	1,750	294
2,000	321	2,000	321
2,250	348	2,250	348

WATER DEMAND ESTIMATES—Continued

Supply Systems Predominantly for Flush Tanks		Supply Systems Predominantly for Flushometers	
Load	Demand	Load	Demand
(Water Supply Fixture Units)	(Gallons per Minute)	(Water Supply Fixture Units)	(Gallons per Minute)
2,500	375	2,500	375
2,750	402	2,750	402
3,000	432	3,000	432
4,000	525	4,000	525
5,000	593	5,000	593
6,000	643	6,000	643
7,000	685	7,000	685
8,000	718	8,000	718
9,000	745	9,000	745
10,000	769	10,000	769

COURTESY NATIONAL STANDARD PLUMBING CODE

SEWAGE FLOWS ACCORDING TO TYPE OF ESTABLISHMENT*

Type of Establishment	Gal. per day per person
Schools (toilet and lavatories only)	15 Gal. per day per person
Schools (with above plus cafeteria)	25 Gal. per day per person
Schools (with above plus cafeteria and showers)	35 Gal. per day per person
Day workers at schools and offices	15 Gal. per day per person
Day Camps	25 Gal. per day per person
Trailer parks or tourist camps (with built-in bath)	50 Gal. per day per person
Trailer parks or tourist camps (with central bathhouse)	35 Gal. per day per person
Work or construction camps	50 Gal. per day per person
Public picnic parks (toilet wastes only)	5 Gal. per day per person
Public picnic parks (bathhouse, showers and flush toilets)	10 Gal. per day per person
Swimming pools and beaches	10 Gal. per day per person
Country Clubs	25 Gal. per locker
Luxury residences and estates	150 Gal. per day per person
Rooming houses	40 Gal. per day per person
Boarding houses	50 Gal. per day per person
Hotels (with connecting baths)	50 Gal. per day per person
Hotels (with private baths—2 persons per room)	100 Gal. per day per person
Boarding Schools	100 Gal. per day per person
Factories (gallons per person per shift—exclusive of industrial wastes)	25 Gal. per day per person
Nursing Homes	75 Gal. per day per person
General Hospitals	150 Gal. per day per person
Public Institutions (other than hospitals)	100 Gal. per day per person
Restaurants (toilet and kitchen wastes per unit of serving capacity)	25 Gal. per day per person
Kitchen wastes from hotels, camps, boarding houses, etc. Serving three meals per day	10 Gal. per day per person
Motels	50 Gal. per bed space
Motels with bath, toilet, and kitchen wastes	60 Gal. per bed space
Drive-in theaters	5 Gal. per car space
Stores	400 Gal. per toilet room
Service stations	10 Gal. per vehicle served
Airports	3–5 Gal. per passenger
Assembly Halls	2 Gal. per seat
Bowling Alleys	75 Gal. per lane

SEWAGE FLOWS ACCORDING TO TYPE OF ESTABLISHMENT*—(Continued)

Type of Establishment	Gal. per day per person
Churches (small)	3–5 Gal. per sanctuary seat
Churches (large with kitchens)	5–7 Gal. per sanctuary seat
Dance Halls	2 Gal. per day per person
Laundries (coin-operated)	400 Gal. per machine
Service Stations	1000 Gal. (First Bay)
	500 Gal. (Each add. Bay)
Subdivisions or individual homes	75 Gal. per day per person
Marinas—Flush toilets	36 Gal. per fixture per hr
Urinals	10 Gal. per fixture per hr
Wash basins	15 Gal. per fixture per hr
Showers	150 Gal. per fixture per hr

***No sewage disposal facilities shall be located on any watersheds of the public water supply system. Privies, septic tanks and underground disposal means shall not be within 200 feet measured horizontally from the high water level in the reservoir on the banks of tributary streams when situated less than 3,000 feet upstream from intake structures. Sewage disposal facilities situated beyond 3,000 feet upstream from intake structures shall be located no less than 100 feet measured horizontally from the high water level in the reservoir or the banks of the tributary streams. Prior to approval, the soil must prove satisfactory by the standard percolation test when underground disposal is used.**

WATER SUPPLY FIXTURE UNITS (WSFU)

Fixture	Number of Fixture Units	
	Private Use	Public Use
Bar sink	1	2
Bathtub (with or without shower)	2	4
Dental unit or cuspidor	—	1
Drinking fountain (each head)	—	1
Hose bibb or sill cock (standard-type)	3	5
House trailer (each)	6	6
Laundry tray or clothes washer (each pair of faucets)	2	4
Service sink	—	4
Lavatory	1	2
Lavatory (dental)	1	1
Lawn sprinklers (standard-type, each head)	1	1
Shower (each head)	2	4
Sink (bar)	1	2
Sink or dishwasher	2	4
Sink (flushing rim, clinic)	—	10
Sink (washup, each set of faucets)	—	2
Sink (washup, circular spray)	—	4
Urinal (pedestal or similar type)	—	10
Urinal (stall)	—	5
Urinal (wall)	—	5
Urinal (flush tank)	—	3
Water closet (flush tank)	3	5
Water closet (flushometer valve)	—	10
Water supply outlet for items not listed above shall be computed at their maximum demand, but in no case less than:		
$\frac{3}{8}$ inch	1	2
$\frac{1}{2}$ inch	2	4
$\frac{3}{4}$ inch	3	6
1 inch	6	10

COURTESY OF MINNESOTA PLUMBING CODE

4

Commonly Used Abbreviations

Certain abbreviations are commonly used in plumbing—on blueprints, job specifications, order catalogs, and other sources of information. The plumber and apprentice must recognize and understand all of these to avoid errors and misinterpretations of data.

ABS acrylonitrile–butadiene–styrene

AGA American Gas Association

AWWA American Water Works Association

BOCA Building Officials Conference of America

B&S bell-and-spigot (cast iron pipe)

BT bathtub

C-to-C center-to-center

CI cast iron

CISP cast-iron soil pipe

CISPI Cast Iron Soil Pipe Institute

CO clean out

CPVC chlorinated polyvinyl choride

CW cold water

DF drinking fountain

DWG drawing

DWV drainage, waste and vent system

EWC electric water cooler

FG finish grade

FPT female pipe thread

FS federal specification

FTG fitting

FU fixture unit

GALV galvanized

GPD gallons per day

GPM gallons per minute

HWH hot water heater

ID inside diameter

IPS iron pipe size

KS kitchen sink

LAV lavatory

LT laundry tray

MAX maximum

MCA Mechanical Contractors Association

MGD million gallons per day

MI malleable iron

MIN (min.) minute or minimum

MPT male pipe thread

MS mild steel

M TYPE lightest type of rigid copper pipe

NAPHCC National Association of Plumbing, Heating, and Cooling Contractors

NBFU National Board of Fire Underwriters

NBS National Bureau of Standards

NPS nominal pipe size

NFPA National Fire Protection Association

OC on center

OD outside diameter

SAN sanitary

SHWR shower

SV service

S & W soil and waste

SS service sink

STD. (std.) standard

VAN vanity

VTR vent through roof

W waste

WC water closet

WH wall hydrant

WM washing machine

XH extra heavy

Index

Abrasive sandcloth, 13
ABS plastic pipe, 73
Adapters, 79-80
  copper, 79
  plastic, 80
Adjustable wrench, 51
A-frame ladder, 26
Air acetylene equipment, 2-5
Air pump (testing equipment), 48
All-purpose wrench, 58
Annealed (soft) copper tubing, 68-69
Appliances
  dishwasher, 116-117, 164
  electric hot water heater, 119-122, 166
  garbage disposal, 117-19, 165
  gas-fired hot water heater, 122-25, 166-67
  washing machine, 125-26, 168
Architect's rule, 40
Architectural drawings, 226
Architectural symbols, 227-29
Area, 200
  circle, 202
  rectangle, 200
  square, 210
  trapezoid, 201
  triangle, 201
Arithmetic, basic, 190-96
Auger bit, 8
Aviator snips, 46

B-tanks, 2
Backwater valve, 112
Ballcock valve, 113
Ball peen hammer, 25
Ball valve, 113
Banging noises in piping, 135
Bars, steel, 6-7
  nail claw, 6
  pry bar, 7
  wrecking bar, 6
Basement leaks, 132
Basin plug wrench, 58
Basin wrench, 53
Bathtubs,
  common problems, 128
  installation, 173
  roughing in, 186
Beaded end cast iron soil pipe, 85
Bell-and-spigot cast iron soil pipe, 86
Benders, copper tubing, 6-7
  geared ratchet lever-type, 6
  lever-type, 7
  spring-type 7
Bituminous pipe, 61-62
Bituminous pipe fittings, 81
Black pipe, 76
Black steel fittings, 81
Box wrench, 53
Brace bits, 8-9
  auger bit, 8
  expansive bit, 9
  screwdriver bit, 9
Brace, standard, 7
Brass fittings, 82-83
Brass pipe, 62-63
Brass reamers, 39
Brushes (for cleaning copper tubing), 13
Bull point chisel, 12
Bushings, 83-84
Butt weld steel pipe, 76

Caps, I.P.S. (testing equipment), 48
Carpenter's level, 29
Cast iron assembly tool, 9
Cast iron soil pipe, 85-88
  beaded end, 85
  bell-and-spigot, 86
  no-hub, 88
Caulking irons, 10-11
  ceiling, 10
  finishing, 10
  inside, 10

long curve, 11
offset, 11
outside, 11
packing, 11
picking, 11
yarning, 11
Ceiling caulking iron, 10
Ceiling leaks, 133
Cesspool, 139-41
common problems, 140
preventive maintenance, 139-40
Chain pipe vise, 37
Chain-type cast iron pipe cutters, 33
Chain wrench, 55
Check valve, 113
Chisels, 11-12
bull point, 11
cold, 12
cutoff, 12
wood, 12
Cleanouts, 90-91
Closet auger, 14
Closet spud wrench, 58
Code, plumbing, 272-73
Coil wires, 14
Cold chisel, 12
Cold water system, 143
Combination square, 47
Combination wrench, 3
Compass saw, 42
Copper adapters, 79
Copper cleaning tools, 12-13
abrasive sandpaper, 13
brushes, 13
copper tubing cleaning tool, 13
Copper fittings, 91-96
compression, 91
D.W.V., 92
flared, 94
pressure, 95
Copper piping
cutting, 158
joining, 158-61
tubes, 69
types of, 67-69
Copper/reaming tools, 39
Copper tubing benders, 6-7
Copper tubing cleaning tool, 13
Couplings, 96-97
CPVC plastic pipe, 73
Crank-operated hand drill, 19
Crosscut saw, 42
Cross fittings, 98
Cutoff chisel, 12
Cutters. *See* Pipe, Cutters
Cutting pipe
brass, 154
cast iron soil, 149
copper, 158
galvanized, 154
plastic, 162
Cylinder wrench, 3

Deburring tools, 38
Decimals and percentages, 194-96
Definitions, 187-190. *See also* Glossary
mathematics, 187-90
plumbing terms. *See* Glossary
Dishwasher, 116-17, 164-65
installation, 164-65
preventive maintenance, 116-117
troubleshooting, 117
Distribution line leaks, 133
Drainage line leaks, 133
Drainage system, 145
Drain cleaning equipment, 14-15
closet auger, 14
coil wires, 14
electric drain cleaner, 15
flat sewer snake, 15
hand-held drain coil, 15
water ram, 15
Drawn (hard) copper tubing, 67
Drill bits (for electric drill), 17
Drill, electric, 16
bits, 17
Drill, hand, 19-20
crank-operated, 19

push, 19
star, 20
Drive sockets, 20-22
flex handle, 21
ratchet-type reversible, 21
speeder, 21
spinner, 21
accessories, 22
Drive socket accessories, 22
Drywall saw, 43
D.W.V.-type copper tubing, 69

Elbows, drainage, 99
Electrical reciprocating saw, 44
Electric drain cleaner, 15
Electric drill. *See* Drill, electric
Electric water heater, 119-22, 166
common problems, 120-22
installation, 166
preventive maintenance, 119-20
troubleshooting, 124-25
Enclosed ratchet manual pipe threader, 34
Expansive bit, 9
Exposed ratchet manual pipe threader, 34
Extension ladder, 26
Extension rule, 40
Extractors. *See* Pipe extractors

Faucets,
kitchen, 169
lavatory centersets, 169
leaks, 133, 180
repairs, 180
Faucet seat wrench, 58
Files, 23
Finishing caulking iron, 10
Fixture leaks, 134
Fixtures, 171-76
bathtubs, 173
kitchen sink, 171
lavatory, 173
roughing-in, 185
vanity-type sinks, 171
Flanges, floor, 100
Flare joints, 159
Flare nut wrench, 53
Flare-Type fittings, 101
Flaring block, 24
Flaring tools, 24-25
flaring block, 24
hammer-type flaring tool, 24
Flashings, roof, 102
Flat sewer snake, 15
Flex handle drive socket, 21
Four-wheel steel pipe cutter, 32
Formulas, math, 211-15
gallons in a tank, 207-10
geometry, 196-206
grade, run, drop, 211-12
offset, 212-15
Fractions, 192-93
Frozen pipes, 141-42
Furnace head (lead joint equipment), 27

Gallons in a tank, calculation, 207-210
Galvanized steel pipe, 76, 156-57
cutting, 155
joining, 157
tapping, 157
threading, 156
reaming, 156
Garbage disposal, 117-19, 165-66
common problems, 118-19
installation, 166-67
preventive maintenance, 117-18
Gas-Fired water heater,
common problems, 123-25
installation, 166-67
preventive maintenance, 122-23
troubleshooting, 122-23
Gasket joints, 153
Gaskets, rubber, 103
Gate valve, 113
Geared ratchet lever-type copper tubing bender, 6

Geometry, 196-206
Glossary, 237-66
Globe valve, 113
Grade, run, drop formula, 211-13

Hacksaw, 43
Hammers, 25-26
- ball peen, 25
- hand-drill, 26
- long-sledge, 26

Hammer-Type flaring tool, 24
Hand-drill hammer, 26
Hand drill. *See* Drill, hand
Hand-held drain coil, 15
Hex wrench, 53
Hose assembly (air-acetylene equipment), 3
Hoses (testing equipment), 48
Hot water heating system, 146
Hot water system, 145
House trap, 104
Hydraulic cast iron pipe cutter, 33

Inside caulking iron, 10
Installation practices, 163-77
- bathtubs, 172
- centrifugal sump pump, 168
- dishwasher, 164
- electric water heater, 166
- fixtures, 171-76
- garbage disposal, 165
- gas-fired water heater, 166
- kitchen faucet, 169
- kitchen sinks, 171
- lavatories, 172
- lavatory centersets, 169
- showers, 175
- submersible sump pump, 169
- toilet, 175
- vanity-type sinks, 171
- washing machine, 168

Internal tubing cutters, 49
Internal wrench, 54
Isometric plumbing drawings, 232-34

Jab saw, 44
Joining pipe, 151-63
- cast iron, 151
- copper, 158-61
- galvanized, 157
- plastic, 162-63
- threaded, 157

Kitchen faucets, 169
Kitchen sinks, 171, 186
- roughing in, 186

K-type copper tubes, 69

Ladders, 26-27
- A-frame, 26
- extension, 26
- step, 27

Lap weld steel pipe, 76
Large tubing cutters, 49
Lavatories, 172
- roughing in, 185

Lead and oakum joints, 151
Lead joint equipment, 27-28
- furnace head, 27-28
- lead ladle, 28
- lead pot, 28
- propane furnace, 28
- rope runner, 28
- shield, 28
- tank key, 28

Lead ladle, 28
Lead pot, 28
Leaks, 131-35
- appliance, 132
- basement, 132
- ceiling, 133
- distribution line, 133
- drainage line, 133
- faucet, 133
- fixture, 134
- pipe and joint, 134

Levels, 29-30
- carpenter's, 29
- line, 29
- plumber's, 29
- torpedo, 30

Lever-type copper tubing benders, 7

Line level, 29
Locking pliers, 38
Long curve caulking iron, 10
Long sledge hammer, 26
L-type copper tubing, 69

Malleable iron fittings, 104
Manual pipe threaders, *See* Pipe threaders, manual
Mathematics
- basic arithmetic, 190-96
- definitions, 187-90
- formulas for plumbers, 207-15
- geometry, 196-206
- symbols, 191
- trigonometry, 207

Maul, *See* Hand-drill hammer.
Measurements
- adding, 219
- converting unlike units, 218-219
- pipes and fittings, 220

Mechanical drawings, 230
Midget tubing cutters, 50
M-type copper tubing, 69

Nail claw, 6
Natural gas piping system, 148
Nipples, 106
No-hub cast iron pipe,
No-hub pipe joints, 154
Noises in piping systems, 135-36
- banging noises, 135
- slamming noises, 136

Nut driver, 30

Offset caulking iron, 11
Offset, hex wrench, 55
Offset regular wrench, 55
Offsets, formulas for, 212-15
Oil can, 31
Oilers, 31
- oil can, 31
- pan oiler, 31
- utility oiler, 31

Open-end wrench, 54
Orangeburg pipe, 61-62
OSHA, 267
Outside caulking iron, 11

Packing caulking iron, 11
Pan oiler, 31
Perimeter, 196-97
- rectangle, 196
- square, 197
- triangle, 197
- circle, 197

Picking caulking iron, 11
Pipe,
- bituminous, 61
- brass, 62
- cast iron, 63-67
- copper, 67-69
- plastic, 69-74
- steel, 74-77
- types of, 61-77

Pipe extractor, 33-34
Pipe fittings, 77-111
Pipe fittings symbols, 234
Pipe cutters, 31-33
- cast iron, 32-33
- steel, 31-32

Pipe measurement, 220-21
Pipe stoppages, 136-37
Pipe taps, 34
Pipe threaders, electric 36
Pipe threaders, manual, 34-35
- enclosed ratchet, 34
- exposed ratchet, 34
- quick opening, 35
- ratchet-type lever, 35
- three-way, 35

Pipe vises, 37
- chain, 37
- yoke, 37

Pipe wrenches. *See* Wrenches, pipe
Plastic adapters, 80
Plastic pipe
- cutting, 162
- joining, 162
- types of, 69-74

Plastic pipe tools, 37
- cutters, 37
- deburring tools, 37

Pliers, 38-39
  locking, 38
  slip joint, 38
  water pump, 39
Plot plan, 225-26
Plumb bob, 39
Plumber's level 30
Plumbing symbols, 231-33
Plugs (testing equipment), 48
Plug wrench, 58
Potable water system, 137-38, 176
Pressure gauge (testing equipment), 48
Pressure, water 138
Propane furnace, 28
Pumps, well 179
Push drill, 19

Quick-acting tubing cutters, 51
Quick-opening manual pipe threader, 35

Radiator wrench, 54
Ratchet-type large manual pipe threader, 35
Ratchet-type reversible drive socket, 21
Reaming galvanized pipe, 156
Reaming tools, 39
  brass and steel pipe reamers, 39
  copper reaming tools, 39
Reciprocating saw. *See* Electrical reciprocating saw.
Regular tubing cutters, 51
Regulations and codes, 272-73
Regulators for B tanks, 4
Relief valve, 120, 121, 122
Repair techniques
  faucet leaks, 180
  pipe leaks, 180
  toilet leaks, 181
  valve leaks, 182
  waste pipe, 182
Repair wrenches. *See* Wrenches, repair
Rope runner, 28
Roughing-in
  basic systems, 184
  bathtubs, 186
  fixtures, 185
  lavatories, 185
  kitchen sinks, 186
  toilet, 185
Ruler, 216-17
  reading, 216
  use according to scale, 217
Rulers and tapes, 40-41
  architect's rule, 40
  extension rule, 40
  steel tape, 41

Safety equipment, 41
Safety rules, 267-71
Saws, hand, 41-44
  compass, 42
  crosscut, 42
  drywall, 43
  hacksaw, 43
  jab, 44
  universal, 44
Saws, power, 44
  electrical reciprocating saw, 44
Schematic drawings, 235
Screwdriver brace bits, 9
Screwdrivers, 45
  special types, 45
Seamless steel pipe, 76
Septic tank, 139-40
Sewage handling system, 138
  common problems, 140
  preventive maintenance of cesspools and septic tanks, 139-40
Sheetrock saw. *See* Drywall saw.
Showers
  common problems, 128
  drains, 127-28
  installation, 175
  preventing problems, 127-28
  stoppages, 183
Shower drains, 127-28
Sinks
  common problems, 128-29

preventing problems, 127
Sink strainer wrench, 54
Slamming noises in piping system, 136
Slip joint pliers, 38
Socket wrench, 58
Soldered joints, 160
Speeder drive socket, 21
Spinner, 21
Spring-type copper tubing bender, 7
Spud wrench, 57
Squares, 47
combination, 47
steel, 47
Squeeze scissor cast iron pipe cutter, 33
Star hand drill, 20
Steam heating system, 148
Steel pipe, 74-77
types of, 76
welds, 76
Steel square, 47
Steel tape, 41
Step ladder, 27
Stillson wrench, 57
Straight hex wrench, 57
Straight snips, 46
Straight wrench, 57
Strainer lock nut wrench, 60
Strap wrench, 57
Striker, 4
Structural drawings, 226
Sump pump
centrifugal, 168
installation of, 168-69
submersible, 169
Symbols
architectural, 227-29
fixtures, 233-38
pipe fittings, 234
plumbing, 231-32

Tank key (lead joint equipment), 28
Tapping galvanized and brass pipe, 156
Taps, 34
Test balls, 49
Testing equipment, 48-49
air pump, 48
caps, 48
hoses, 48
plugs, 48
pressure gauge, 48
test balls, 49
testing plug, 49
Testing plug, 49
Threaders. *See* Pipe threaders, electric; Pipe threaders, manual
Threading brass and galvanized pipe, 157
Three-way manual pipe threader, 36
Toilet
common problems, 130-31
installation, 175
leaks, 181
preventing problems, 129-30
roughing in, 185
stoppages, 131, 183
Torch handle, 4
Torch tips, 4
Torpedo level, 30
Torque wrench, 54
Trap, 110
Trigonometry, 207
Tubing cutters, 49-51
internal, 49
large, 49
midget, 50
quick-acting, 51
regular, 51

Unfreezing pipes, 142
Unions, 111
Universal saw, 44
Utility oiler, 31
Utility wrench, 60

Valves, 111-12, 182
backwater, 111
ball, 112
ballcock, 112

check, 112
gate, 112
globe, 112
leaks, 182
Vanity-Type sinks, 171
Vise grip pliers, *See* Locking pliers
Vises. *See* Pipe vises
Volume, 202-206
cube, 203
cylinder, 206
rectangle, 202
trapezoidal prism, 203
triangular prism, 203

Washing machine, 125-26, 168
common problems, 126
installation of, 168
preventive maintenance, 126
Waste pipe stoppages, 182
Water closet. *See* Toilet
Water heater, electric, 119-22, 166
common problems, 120-22
installation, 166
preventive maintenance, 119-20
troubleshooting, 124-25
Water heater, gas-fired, 122-25, 166-67
common problems, 123-25
installation, 166-67
preventive maintenance, 122-23
troubleshooting, 124-25
Water key, curb wrench, 54
Water main service, 177
Water meter, 145
Water, potable, 176-69
well, 177
fittings, 179
Water pressure problems, 138
Water pump pliers, 39
Water ram, 15
Water softener, 138
Weather-caused problems, 141-42
preventing frozen pipes, 141
preventive maintenance, 141
unfreezing pipes, 142
Wells, 177-79
lines from, 179
types of, 178-79
Wheeled electric pipe threader, 36
Wide roll steel pipe cutter, 32
Wood chisel, 12
Wrecking bar, 6
Wrenches, general purpose, 51-54
adjustable, 51
basin, 53
box, 53
combination, 53
flare-nut, 53
hex, 53
internal, 54
open-end, 54
radiator, 54
sink strainer, 54
torque, 54
water key, curb, 54
Wrenches, pipe, 55-57
chain, 55
end, 55
offset, hex, 55
offset, regular, 55
spud, 57
strap, 57
stillson, 57
straight, 57
straight hex, 57
Wrenches, repair, 57-60
all-purpose, 58
basin plug, 58
closet spud, 58
faucet seat, 58
plug, 58
socket, 58
strainer lock nut, 60
utility, 60

Yarning caulking iron, 11
Yoke pipe vise, 37